Chemometrics:
A Practical Guide

WILEY-INTERSCIENCE SERIES ON LABORATORY AUTOMATION

ADVISORY BOARD

Chemometrics: A Practical Guide

KENNETH R. BEEBE
RANDY J. PELL
MARY BETH SEASHOLTZ
The Dow Chemical Company

A Wiley-Interscience Publication
JOHN WILEY & SONS, INC.

New York · Chichester · Weinheim · Brisbane · Singapore · Toronto

Library of Congress Cataloging-in-Publication Data:

Beebe, Kenneth, R.
      Chemometrics : a practical guide / Kenneth R. Beebe, Randy J. Pell, Mary Beth Seasholtz.
          p.   cm. — (Wiley-Interscience series on laboratory automation)
      "A Wiley-Interscience publication."
      Includes index.
      ISBN 0-471-12451-6 (alk. paper)
      1. Chemistry, Analytic—Statistical methods.   2. Chemistry, Analytic—Mathematics.   I. Pell, Randy J.   II. Seasholtz, Mary Beth.   III. Title.   IV. Series.
QD75.4.S8B44   1998
543'.001'5195—dc21

Printed in the United States of America.

10 9 8 7 6 5 4

To Sue, Lauren, and Emily Beebe, to Karen Pell, and to Jon Zieman— the families of the authors—who gave up many hours of companionship during the preparation of this book. Without their support and encouragement, this manuscript would have gone the way of so many other good intentions.

Contents

neutral network

Pattern Reg :

Matlab - chem tool box

Mathworks

time series analysis
signal analysis

PLS Toolbox (free)
   a) public domain version PLS Toolbox vers 1.3 / Scout
      source code - Eigenvector Technolgis, maker of toolbox   PLS
   b) new version $600. nice 3D graphis cap

OcTAVE: Matlab clone. GNU family Run on UNIX/Linux

Preface

Our goal in writing this book was to develop a practical guide for problem solving using the most widely available chemometric tools. It is written from the point of view of an analyst using the tools to solve a problem. Most of the text concentrates on the interpretation of the diagnostics that are encountered when using a particular pattern recognition or multivariate calibration method. Enough theory is included to help the reader understand how the methods work in order to insure that the techniques are appropriately applied. The intent was to take the middle ground between a cookbook and a purely theoretical approach. The reader is referred to software manuals for the details of how specific software is used and to the literature for in-depth technical descriptions of the methods.

The format of this book grew out of our experiences in applying chemometrics to solving problems in an industrial setting. Throughout the years, we have done this either as members of teams or as coaches teaching co-workers the use of chemometric tools. In both roles, it is important to be able to define and solve the problem and effectively communicate the solution. This book documents our process for applying the most widely used pattern recognition and multivariate calibration techniques. In this process we identify six steps, or "habits," that we believe lead to reliable results when they are consistently applied.

During various phases of preparing this manuscript, we gathered feedback from colleagues on content and form. We are grateful to the following individuals for their insightful reviews of various portions of the manuscript: Marlana Blackburn, Stan Deming, Terry Kram, Bryant LaFreniere, Tormod Næs, Karen Pell, Scott Ramos, Sonja Sekulic, Ken Stutts, Fred Van Damme and Yongdong Wang.

We also owe a debt of gratitude to our parents—Fumiko Beebe, Glen and Darlean Pell, and Arlin and Dalene Seasholtz—for their encouragement in our lives and academic pursuits.

1

The Six Habits of an Effective Chemometrician

We are what we repeatedly do. Excellence, then, is not an act, but a habit.

—Aristotle

The goal in writing this book was to develop a practical guide for problem solving using the most widely available chemometric tools. The methods are described using a systematic flow that follows what we term the "Six Habits of an Effective Chemometrician." These habits are the steps that must be followed in order to consistently develop reliable models. In his book on leadership, Covey highlights the need for constancy of action when he uses the Aristotle quote cited above to introduce his seven habits of effective people (Covey, 1989). Likewise, the chemometric habits are useful for insuring excellence in data analysis when they are consistently applied.

Before describing the six habits, it is important to define what is meant by the term chemometrics. A general definition is "the use of statistical and mathematical techniques to analyze chemical data." In this book, we prefer the broader definition of chemometrics as "the entire process whereby data (e.g., numbers in a table) are transformed into information used for decision making."

To affect this transformation, the first question to pose is "What do we want to know about the system?" Spending time defining the system and problem helps insure that appropriate measurement techniques are selected and protocols are developed to gather relevant data about the system. This "Defining the Problem" step involves skills and knowledge about the chemistry of the system, the required precision and accuracy of the determinations, the measurement system, the available chemometric methods of analysis, and experimental design methodologies (see Chapter 2). The chemometrician should be involved at this early stage of investigation to insure that the mathematical requirements of the analysis are understood as well as any special requirements for the chemometric tool being considered.

It is only when the performance objectives of the project have been defined that the process of collecting the data should begin. Once collected, chemometric tools can be used to transform the data into useful information. This third step, "Data Analysis," is where the chemometric tools described in Chapters 3–5 are applied. Deriving the best information is often accomplished using an iterative approach where the chemometrician analyzes the data, discusses the preliminary results with the chemist/engineer, and refines the model accordingly. In this way, chemical knowledge is coupled with mathematical knowledge and the synergism results in better understanding and ultimately better decisions.

To appreciate the niche that chemometric methods fill in problem solving, it is instructive to discuss the difference between explicit and implicit modeling approaches. Explicit modeling is used when a system can be described by an accepted law of chemistry or physics. Predicting concentration based on Beer's law and an absorbance measurement is an example of the use of an explicit model (absorbance = absorptivity * pathlength * concentration). When the explicit mathematical relationship is known, problems are often most efficiently solved using the explicit models. This is because it generally takes less effort to estimate the parameters of the model. In effect, one is taking advantage of the effort that has already been expended by other researchers to validate the physical model form.

In contrast to explicit models, implicit modeling uses empirical models to extract useful information from data without assuming a physical description of the system. Fitting a nonlinear equation to a table of data to create an online look-up table is an example of an implicit model. The nonlinear equation describes the entries in the table in that it is able to reproduce the values without having a basis in physics. The implicit approach is often the most effective when the analyst is faced with complex data and no known model. However, because implicit models have no fixed form, it is very important to validate the reliability of the model.

Although the chemometric tools discussed in this book fall into both of these categories, most of the emphasis is on implicit modeling. Valid explicit models are not common in practice and, therefore, implicit models are often necessary to analyze the data and/or construct predictive models.

1.1 USING THIS BOOK

The chapters of this book outline a recommended process for solving problems. Chapter 2 begins with defining the problem and designing the experiments and includes some helpful tools for effective consulting. Chapter 3 discusses preprocessing the data to prepare for the data analysis phase. Chapters 4 and 5 present the tools that are used to transform the data into useful information (Pattern Recognition and Multivariate Calibration/Prediction, respectively).

Chapters 4 and 5 each begin with an introduction that discusses the general theory and uses for the tools. This is followed by a decision tree (flow diagram) that aids the reader in selecting the specific tool to use for a given problem. For each method discussed, examples are presented to illustrate how the tools are used. The "Example 1" section uses a simple data set where the "truth" is generally known. When presenting results for this example, the diagnostics for the methods are discussed in detail to explain how the output is interpreted. This includes discussion on the utility of the different output diagnostics and what can go wrong with an analysis (often using "what if" scenarios). For each method, at least one additional example is presented where the method is applied to a real-world problem. This section does not include as much teaching as in Example 1. The intent is to teach the how and why in Example 1 and then see the tools in action in subsequent examples.

1.2 THE SIX HABITS OF AN EFFECTIVE CHEMOMETRICIAN

Each of the tools in Chapters 4 and 5 are presented in the context of the "Six Habits of an Effective Chemometrician." These habits provide a recipe for systematic evaluation of the data regardless of the method being used.

Habit 1. Examine the Data
Once an appropriate data set has been collected, the first step is always to examine the data. This is usually accomplished by examination of plots and/or tables. The primary purpose of this step is to use the human eye to look for "obvious" errors or features in the data. Because errors can occur in both the measurement variables (e.g., spectra) or characteristic values (e.g., concentrations), it is important to examine both of these sets of numbers in this step. This initial view of the data may indicate the need for preprocessing, highlight features, or samples in the data set that warrant further investigation.

Habit 2. Preprocess as Needed
There can be random or systematic sources of variation that mask the variation of interest. This unwanted variation may reduce the effectiveness of the model. An understanding of the chemistry or physics underlying these unwanted sources of variation helps with appropriate selection of preprocessing techniques (see Chapter 3).

It is important to remember that preprocessing changes the data set and, if inappropriately applied, can remove important variation from the data. Therefore, it is necessary to reexamine the data set (Habit 1) after preprocessing has been applied to insure that it has accomplished what was expected. For the examples presented in this book, the authors were careful to reexamine data where preprocessing was applied. However, the plots of the transformed data are presented only when they have instructional value.

Habit 3. Estimate the Model

The next step is to generate the chemometric model and associated diagnostics. Depending on the software used, some initial user input may be required before the calculation takes place. These required inputs are only briefly described in Habit 3 because they are often software specific.

The output results from different software packages also differ. We have included the most common diagnostics in this book, but it is important to consult the software user's manual to understand any differences that may be encountered. There may also be additional diagnostic tools in specific software packages that are not presented in this book. The reader is encouraged to take full advantage of these additional tools.

Habit 4. Examine the Results / Validate the Model

All of the chemometric methods generate numerical and graphical results. Habit 4 describes how to examine the computer output with the goal of validating the model (i.e., determine if the model is reliable). Diagnostic tools for each of the methods are used to assess the confidence that can be placed in the results. If the model is not acceptable, refinement is often possible within Habit 4 by adjusting the model parameters.

Beware that building models is trivial; a computer program will almost always produce a model unless there is a drastic problem with the data. Building reliable models is more difficult. The computer does not automatically assess the reliability of the results and therefore chemical and mathematical knowledge must be used to determine whether the model is reasonable.

For reference, each of the Habit 4 sections concludes with a summary table discussing all of the diagnostic tools presented for the method. The first column in the table lists the name of each diagnostic and the second column describes the output of the diagnostic including: (1) what it is used for, (2) a description of what to expect from well-behaved and problematic data, and (3) any special features of the diagnostic. The tools are also grouped into three categories (model, sample, and variable) according to their use. The primary use of the model diagnostic tools is to determine the quality of the model and to select model parameters (e.g., the number of latent variables to include in a PLS model). The sample diagnostic tools are used to study the relationships between the samples and to identify unusual samples. The variable diagnostic tools do the same, but for the variables.

Habit 5. Use the Model for Prediction

For the methods that generate a predictive model, Habit 5 is the application of the model to an unknown sample(s). The output from this habit is the predicted properties or classes of the unknown samples. For example, with PLS, calibration models are constructed to predict properties of future samples.

Habit 6. Validate the Prediction

The computer rarely fails to produce a prediction result given a model and an unknown. It is therefore important that even apparently reasonable results be validated. Being able to validate prediction results is one of the greatest advantages of using multivariate techniques. With the prediction diagnostic tools, it is possible to determine when the model is not applicable because of instrument failure or unusual unknown samples. Validation increases the chances of making good decisions based on the outputs from the models by indicating the confidence that should be placed on the predicted values. As with the tools discussed for Habit 4, the prediction diagnostic tools are summarized in a table for easy reference.

1.3 DEFINITIONS

It is important to define some terms so that the reader will understand their usage in this book as well as in the chemometrics literature. The following are words that are used throughout the book and their definitions.

Analyte or Component   A chemical species contained within a chemical sample. For example, iron may be present in water samples. The goal of a chemical analysis is often to determine the concentration of analytes in samples.

Calibration   The process of constructing a model that is used to predict characteristics or properties of unknown samples. The model is constructed from a calibration data set with measured multivariate responses (**R**) and corresponding known sample concentrations or physical characteristics of interest (**C**).

Calibration or Training Set   The collection of samples that is used to construct a calibration or classification model. (*See also* Validation Set.)

Concentration, Characteristic, or Property Matrix (C)   Different characteristics or properties of chemical systems can be predicted using multivariate calibration techniques. Throughout this book, the word concentration is often used as a generic term to represent concentration, characteristic, or property.

Cross-Validation   A process used to validate models whereby the calibration set is divided into calibration and validation subsets. A model is built with the calibration subset and is used to predict the validation subset. This process is repeated using different subsets until every sample has been included in one validation subset. The predicted results are then used to validate the performance of the model.

Factor   The result of a transformation of a data matrix where the goal is to reduce the dimensionality of the data set. Estimating factors is necessary to construct principal component regression and partial least-squares models, as discussed in Section 5.3.2. (*See also* Principal Component.)

Fit versus Prediction   A distinction is made between model fit and model prediction. Fit is a measure of how well the model describes the data used to construct the model. This can be a misleading measure of the predictive ability when using an implicit model because the model can be manipulated to achieve a good fit even when a causal relationship does not exist. For this reason, many of the diagnostic tools discussed in Habit 4 use a prediction criterion based on cross-validation or separate prediction sets to evaluate model predictive ability. (*See also* Overfitting.)

Matrix Notation   Although an understanding of linear algebra is not necessary to use this book, equations and mathematics are included where they add to the understanding of a concept or aid the reader when referring to other technical works. Where equations are used, the following font types are employed to represent the corresponding data types:

nfact—the number of factors used to construct a model

npure—the number of pure components

nsamp—the number of samples

nvars—the number of measurement variables

scalars (single numbers)—italic lowercase letters (e.g., x)

vectors (a row or column of scalars)—bold lowercase letters (e.g., $\mathbf{x}$)

matrices (a table of numbers)—bold uppercase letters (e.g., $\mathbf{X}$)

transpose of a vector or matrix—superscript T (e.g., $\mathbf{X}^T$ is the transpose of $\mathbf{X}$)

pseudoinverse of a matrix—superscript † (e.g., $\mathbf{X}^\dagger$ is the pseudoinverse of $\mathbf{X}$)

Measurement Variable   A single reading (scalar) made on a sample. For example, the turbidity of a wastewater stream sample is a measurement variable. (In some pattern-recognition literature, the term feature is used in place of variable.)

Measurement Vector   A term referring to a collection of measurement variables for a single sample. For example, the near-infrared spectrum of a sample is a measurement vector. (*See also* Spectrum.)

Multivariate   A multivariate measurement is defined as one in which multiple measurements are made on a sample of interest. That is, more than one variable or response is measured for each sample. Using a sensor array to obtain multiple responses on a vapor sample is a multivariate measurement.

Overfitting   Overfitting occurs when the model used to describe a data set is overly complex. An example of this in regression analysis is the use of a second-order polynomial to describe the relationship between two variables when the true relationship is a straight line. In chemometrics, the most common example of overfitting is the use of too many latent variables in a partial least-squares model. Avoiding overfitting is one of the primary purposes of the validation step in the calibration process. (*See also* Underfitting.)

Pattern Recognition A process of examining the relationships between samples and/or variables in a data set. Unsupervised pattern-recognition tools are used to determine if there are groupings of similar samples in a data set. Supervised pattern-recognition tools are used to classify unknown samples as more likely of type A or type B.

Prediction Prediction is the process of estimating the characteristics of unknown samples by applying a calibration model to an unknown measurement vector. (*See also* Calibration.)

Principal Component (PC) In this book, the term principal component is used as a generic term to indicate a factor or dimension when using SIMCA, principal components analysis, or principal components regression. Using this terminology, there are scores and loadings associated with a given PC. (*See also* Factor.)

Residuals A measurement residual is the part of a measurement vector that is not explained by a model (Residual = Actual Data − Data Reconstructed Using the Model). For example, when using a factor-based method to build models with spectral data, the portion of a spectrum that is not used by the model is the residual. Concentration residuals are the differences between the true (or known) and predicted concentrations. These are only available at method validation.

Response Matrix (R) Multivariate calibration and pattern-recognition techniques make use of multiple responses that are represented by the matrix **R**. The rows of **R** contain the measurement vectors for individual samples. This matrix has the dimensions nsamp × nvars.

Sample The term sample refers to a single chemical sample as is understood by a chemist. This is in contrast to the statistical definition of a sample, which is a collection of observations from a population.

Spectrum or Spectra To simplify the discussions, the term spectrum or spectra is used in this book as a generic term to refer to a collection of measurements on a sample (i.e., a measurement vector). The reader should not infer from the use of this term that the methods can only be applied to spectroscopic data. (*See also* Measurement Vector.)

Underfitting Underfitting occurs when the model used to describe a data set is too simple. An example of this in regression analysis is the use of a straight line to describe the relationship between two variables when the true relationship is quadratic. (*See also* Overfitting.)

Validation or Test Set A set of samples used to validate a prediction or classification model. These samples are not part of the calibration set that is used to construct the model. (*See also* Calibration or Training Set and Cross-Validation.)

1.4   SUMMARY OF THE SIX HABITS OF AN EFFECTIVE CHEMOMETRICIAN

The goal in writing this book was to develop a practical guide for solving problems using chemometrics and to provide a "ready reference" for the widely accepted, understood, and used methods. The "Six Habits of an Effective Chemometrician" are presented as a recipe for systematic development of models and evaluation of data regardless of the method employed. Following these steps and making them habits in your work will allow you to make the most of the tools found in the remainder of this book.

REFERENCE

S. R. Covey, *The Seven Habits of Highly Effective People: Restoring the Character Ethic,* Simon & Schuster, New York, 1989.

CHAPTER

2

Defining the Problem

Kotuna ejali mabe te. (To ask is not bad.)

—Old Lingala Saying (Zaire)

Taking the time to adequately define the problem before working on a solution helps insure that the right problem is solved in an efficient manner. This problem definition includes asking the right questions, listening to the answers, framing the problem in chemometric terms, and planning for how the problem will be solved.

Much of the discussion in this chapter is from the point of view of a consultant (chemometrician) working with a project sponsor. However, the concepts discussed here can also be applied if the chemometrician is an independent researcher. As a consultant, the chemometrician aids in defining the problem, planning and executing experiments, and analyzing the data.

Consulting skills are valuable in any situation where a team is working together to solve a problem. Each team member brings a different perspective and knowledge base that should be exploited. The team's knowledge base is used to gain a mutual understanding of the system under investigation. The key to effectively facilitating this process is good communication. This may require that different team members learn the basics of experimental design, measurement science, and/or chemometrics. Likewise, the chemometrician often must understand the impact of engineering constraints and business needs in order for the project to be successful. When everyone has an understanding of the entire project, the team can work together to define the correct problem. Not taking the time to adequately define the problem before deciding on a solution often leads to unsatisfactory results. When this happens, it becomes clear that the Lingala saying "To ask is not bad" is better rephrased "To NOT ask IS bad."

2.1 DECISION TREE

The basic steps that are taken to solve an analytical problem can be described using the scientific problem-solving approach shown in Figure 2.1. The first step in defining the problem is gaining an understanding of the context of the

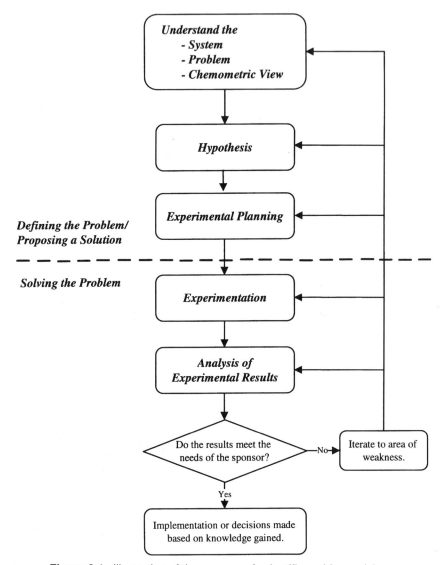

Figure 2.1. Illustration of the process of scientific problem solving.

problem (i.e., the system). Once the system is understood, it is possible to frame the problem that is to be solved. At the same time, possible chemometric approaches to the problem are considered to insure that data are collected that are appropriate to solve the problem.

This understanding of the problem leads to an assumption or "hypothesis" being posed regarding the system. The hypothesis is the problem definition and a proposed solution which includes consideration of how possible chemometric tools can be used to solve the problem. The hypothesis determines how the system is to be probed and leads to a plan for collecting the data. This "experimental planning" is the last step in the problem definition phase.

It is important to understand that all of these steps equally impact the success of the project. Not understanding the system can lead to a problem definition that when solved does not help improve the system. An unreasonable hypothesis is by definition defining the wrong problem. Poorly designed experiments and faulty experimental technique leads to bad data. And improper data analysis can transform even good data into useless results. Careful attention to all steps in this process is therefore required in order to achieve optimal results.

The following sections discuss each of these steps in detail and present a process that can be used to help insure successful problem definition. The two steps in the "Solving the Problem" phase of the project are not discussed in this chapter. "Experimentation" is dependent on the method that is used to probe the system and relies on the expertise of the person(s) collecting the data. "Analysis of Experimental Results" is the application of the chemometric tools and is the topic of Chapters 3-5.

2.2 COMMUNICATION, COMMUNICATION, COMMUNICATION

One of the keys to solving the right problem is to define it correctly in the first place. To help insure that this happens, it is recommended that the chemometrician be involved early in the project. There are three phases of communication for effective problem definition: (1) asking, (2) listening, and (3) feedback. These phases are not necessarily serial and are most likely repeated many times during the course of the project. "Asking" effective questions focuses the discussion. "Listening" allows the chemometrician to understand what the sponsor wants to know about the system. "Feedback" is interaction between all parties that leads to a deeper understanding of the system and the problem.

2.2.1 Understanding the System

The first objective of communication is *not* to understand the *problem,* but to understand the *system.* The system includes everything within a set of boundaries and everything that crosses those boundaries (i.e., inputs and outputs). Problems must be framed in the context of a system, and one who is ignorant of the system will have difficulty formulating a reasonable solution. Gaining this knowledge is best achieved by asking the sponsor many questions. A good rule of thumb is to apply the late Tomas Hirschfeld's advice: "Ask questions until they stutter." It is also important to ask the same question in several different ways so as to gain a fuller appreciation of the system.

The sponsor is asked to describe the inputs, the processes that take place within the system boundaries, and the outputs. In the future, a goal will be to devise measurement systems to probe one or all parts of the system. A graphical representation of this general model is shown in Figure 2.2.

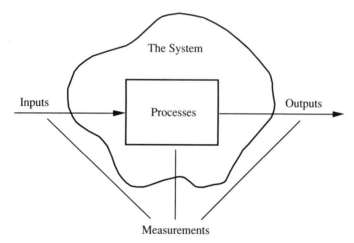

Figure 2.2. Simplified view of a chemical system under investigation.

Some generic questions that are useful in defining the system are found in Table 2.1. These have been categorized into seven groups: preliminary, historical, safety, inputs, process, outputs, and wrap-up. The preliminary questions help to define the team and the scope of the project. Historical questions are asked to gather existing information about the system. Safety concerns are always an important consideration. The input, process, and output questions are asked to gather more detailed information about the system.

The list of questions in Table 2.1 is not exhaustive, but is useful as an outline for discussion. Remember that the initial goal is to understand the system before trying to solve the problem. Specific questions may be added depending on the setting of the system.

A list of additional questions pertaining to process analytical chemistry is found in Table 2.2. For example, it is instructive for the sponsor to present a block diagram of the entire production process. This is often enlightening even if the proposed project only involves one unit operation in a large plant. The other operations are either inputs or outputs to that system, and therefore must be considered. This discussion also educates everyone in the terminology used in the plant and the different engineering and business constraints that are present.

2.2.2 Understanding the Problem

The goal of understanding the problem is to formulate an hypothesis (step 2 in Figure 2.1). The problem resides in the domain of the system, and this is why it is critical to examine the system first. When the effort is expended to understand the system, it will be apparent how solving the problem impacts the system. Again the key is listening to the sponsor and asking appropriate questions. Now, however, the focus is shifted from the system to the problem.

TABLE 2.1. Questions that are Helpful for System Definition

Generic System Definition Questions

Preliminary

Why is there interest in working on this project?

What kind of a priority is this project for the sponsor?

Is it a feasibility study?

Who is responsible for what?

Who is on the team?

What is the structure of the team?

Who understands the chemistry in detail?

Are there specific regulatory constraints?

Historical

Have others worked on this system before?

If they were not successful, why did they fail?

What does the literature say about the system?

Safety

Are there any hazards to consider?

Is anything particularly toxic?

Inputs

Describe all of the inputs to the system (chemical and energy).

How are the inputs added, measured and controlled?

Process

Describe the chemical reaction(s).

What measurements are presently made during the reaction(s)?

What does the flow diagram of the system look like?

Outputs

Describe the outputs of the process (chemical and energy).

What measurements are presently made on the outputs?

Wrap-up

What questions have I not asked that I should have asked?

Now, tell me about your problem.

Table 2.3 contains a list of questions that can be used to gain this perspective. These questions are intended to stimulate a brainstorming discussion to understand the problem, not to solve it.

Once perspective is gained on the problem, the view is narrowed by framing it in mathematical measures of performance. The second section of Table 2.3 lists some questions that can be asked. It is important to define the required accuracy and precision of the analysis and insure that everyone appreciates the distinction between these two concepts. Also, if accuracy is going to be evaluated, there must be agreement on a standard of comparison (a reference method).

During the course of defining the problem, one or more techniques will emerge as being suitable. At this point the chemometrician should begin asking technique-specific questions to determine which approach is most

TABLE 2.2. Process Analytical Chemistry Specific System Definition Questions

Supplemental Process Analytical Chemistry System Questions

Preliminary
    Who are the plant contacts?
    Who are the R&D contacts?
    Who understands the process in detail?
    What funds are available for capital?
    What is the value of this project to the business?
    Is a block diagram of the process available?
Safety
    What is the classification and division of the area?
    Are there special line-opening considerations?
    Where are the Material Safety Data Sheets?
    What protective equipment is needed?
    What materials of construction are appropriate?
Inputs
    What are the raw materials and what phases are they?
    What impurities can be present in the raw materials?
    Who supplies the raw materials?
    What are the specifications on the raw materials?
    How are the specifications currently measured?
    What are the pressures and temperatures of the inputs?
Process
    How do you currently control the process?
    What is the process residence time?
    Is the process batch or continuous?
    Is there a recycle stream? If so, how is it handled?
    Where are the current sampling points?
    Where are the most useful sampling points?
    How many phases are present at each sampling point?
    How representative is the sample?
    What are the pressures and temperatures in the process?
    Describe the chemistry during normal operation and product transition.
    Is there a model of the process (kinetic, mass balance, etc.)?
Outputs
    How is the product packaged?
    Who is the customer for the final product?
    What are the specifications for the product?
    How are the specifications currently measured?
    Where are the sampling points?
    Where are the most useful sampling points?
    How many phases are present at each sampling point?
    What are the pressures and temperatures in the output?

TABLE 2.3. Questions that are Helpful for Defining the Problem

Generic Problem Definition Questions

Defining the problem in terms of the system
  What is the goal; how will success be measured?
  Where in the system do you think the problem is located?
  Can the question be answered with existing data?
  What are your ideas on the solution?
  What needs to be measured?
  What is the minimum data to answer the question?
  What are the time and resource constraints on the data collection?
  How does solving the problem improve the system?
  What is the decision that needs to be made?
  What is the cost of being wrong (or right)?
  Are there regulatory constraints on the solution?
Mathematical measures of performance
  What is the required precision? Why?
  What is the required accuracy? Why?
  What reference method will be used to define the truth?
  What is the measurement error in the reference method?
Chemometric and measurement specific (Not an exhaustive listing!!)
  Can you obtain pure spectra?
  What is the phase(s) of the sample?
  How fast do you need the analysis results? What is the basis of this need?
  What is the required detection limit?
  What are the expected interferences, number, and ranges?
  How will the model be updated or recalibrated?

appropriate. At the same time, it is also important to gather information to determine which measurement technique will be used (e.g., NIR vs. NMR vs. GC/MS). The last section in Table 2.3 lists some chemometric and measurement-specific questions. As an example, if it appears that classical least squares (see Section 5.2) is a reasonable approach, it is germane to ask whether the pure or mixture spectra can be obtained.

Although it is necessary to propose potential solutions during these discussions, beware of settling on a final path too soon. Priorities change, unexpected events occur, and best guesses are sometimes off the mark. For this reason, it is important to keep as many options open as is possible.

Only a small subset of method-specific questions are shown in Table 2.3 because of the breadth of chemometric and measurement methodology. Understanding the information in Chapters 3–5 will help generate chemometric-related questions. The decision trees in these chapters can also help determine appropriate method(s) for a given situation. In time, your experience will also become a valuable resource.

2.2.3   Understanding the Chemometrics View

Communicating the chemometric view is important for at least two reasons. First, there is often a cultural barrier between chemometricians and their sponsors. It is therefore important that the chemometrician build a bridge between mathematics, chemistry, and engineering. Giving the sponsor a quick tutorial on a method is useful to demonstrate that the proposed approach is reasonable. It is also important to explain the limitations of the method(s). This exhibits an understanding of the tools and explains why the methods have certain requirements. For example, a very powerful advantage of inverse modeling techniques is the ability to account for the presence of impurities (see Section 5.3). However, the sponsor must understand that the experiments must be designed properly to realize this advantage.

The second reason for describing the proposed chemometric solution is to perform a "reality check." In understanding the requirements for the tool, the sponsor may point out a limitation that had not been considered. For example, assume for an on-line application that NIR with direct classical least squares has been selected as the best solution. This approach has a requirement that the pure spectra are available. When this requirement is made clear, it may be determined that they are not available at the proposed sampling point. With this new information, multiple sampling points with fiber optics or a different chemometric method must be considered. This example illustrates how an understanding of the chemometric view enables the sponsor to contribute to the chemometric solution.

2.3   EXPERIMENTAL PLANNING

Planning the experiments is the next step after the hypothesis has been formulated and a solution has been proposed (see Figure 2.1). This includes planning what data to collect and how it is collected. Specifically, this means determining: the number, identity, and order of samples to analyze, what is to be measured, what is to be held constant, how the data will be collected, and any other factors that affect the data that are generated.

It should be noted that we make a distinction between experimental *planning* and *design.* In this chapter, experimental designs are defined as specific statistical constructs such as factorial or central composite designs (Box et al., 1978). Experimental planning encompasses everything that needs to be considered in choosing experiments. The design is the outcome of the planning.

It is clear that all experiments follow a design—either the data collection phase is consciously planned or fate will run its course. This is critical to realize because the experiments have a direct impact on the results. For example, it is not possible to estimate a slope and intercept with only one data point (a minimum of two are required). Another simple example of poor planning is to calibrate a viscometer without compensating for varying temperature. Because viscosity depends on temperature, this information is needed to construct the calibration model. These examples show how the data that have been collected limit the model(s) that can be estimated. Keep in mind this interdependence between the data collected and the proposed model(s) when planning experiments.

2.3.1 Benefits of Experimental Planning and Design

In this section, three benefits of planned and designed experiments are discussed.

1. *Efficient Use of Time and Resources.* The probability of collecting useful data is increased by taking time to consider what can affect the results and ultimately what experiments to run. Designed experiments use resources efficiently to maximize the knowledge gained from the experiments.

2. *Designed Experiments Produce More Precise Models.* In the context of linear regression, this is demonstrated by examining the statistical uncertainties of the regression coefficients. Equation 2.1 is the regression model where the response for the ith sample (r_i) of an instrument is shown as a linear function of the sample concentration (c_i) with measurement error (e_i),

$$r_i = b_0 + b_1 c_i + e_i \tag{2.1}$$

where i ranges from 1 to the number of samples in the design (nsamp). Assuming the linear equation adequately models the system, the objective is to determine precise estimates of the slope (b_1) and intercept (b_0). In statistical terms, this is achieved by estimating unbiased coefficients with minimum uncertainties. The equations for the estimated variances (i.e., uncertainties) for the slope $[\text{var}(b_1)]$ and intercept $[\text{var}(b_0)]$ are shown in Equations 2.2 and 2.3, respectively (Draper and Smith, 1981),

$$\text{var}(b_1) = \frac{s^2}{\sum_{i=1}^{\text{nsamp}} (c_i - \bar{c})^2} \tag{2.2}$$

$$\text{var}(b_0) = \frac{\left(\sum_{i=1}^{\text{nsamp}} c_i^2\right) s^2}{\text{nsamp} \sum_{i=1}^{\text{nsamp}} (c_i - \bar{c})^2} \tag{2.3}$$

where s^2 is the regression residual sum of squares and $\bar{c}$ is the average of the concentrations. One way of decreasing both of the variances is by increasing the range of the concentrations (i.e., increasing $c_i - \bar{c}$). Similarly, the variance of the intercept can be decreased by increasing the number of samples in the design (i.e., nsamp). Finally, everything else being equal, minimizing the c_i^2 will result in a smaller variance of the intercept. All these terms ($c_i - \bar{c}$, nsamp, c_i^2) are determined by the experimental design before any data are collected.

3. *Helping to Find Models that Reflect the Truth.* Careful planning before collecting data helps insure that important factors are considered when studying the system. This increases the chances that the system will be correctly characterized. In addition, good experimental designs can help determine when a postulated model is inadequate for describing the system. Data from designed experiments allow for a comparison between the fit of the data to the model and the expected natural variation in the measurements. In statistical terms, this is known as examining the lack of fit. In indicating lack of fit, the data may not always reveal *why* the model is inadequate, but at least the analyst knows that the work is not completed.

2.3.2 Steps for Designing the Experiment

Assuming that an initial problem specification has been established, the next step is to determine what data are needed to solve the problem. This is where the concepts of experimental designs are extremely useful. The discussion that follows incorporates the more classical concepts of experimental design into a chemometric context. An example is given that demonstrates why it is necessary to blend the more classical approaches and chemometric practices to construct a viable design. The reader is referred to the literature for more in-depth discussion on the classical approaches (Box et al., 1978; Deming and Morgan, 1987).

Step 1—Identify the Variables

Generating an experimental design begins with identifying all variables that impact the system under investigation. The objective is to revisit the system after a solution to the problem has been proposed. The scope of the discussion should remain broad, so as not to overlook significant variables. A brainstorming session with the sponsor can help insure that a comprehensive list of variables is generated.

Although the following discussion centers around the type of multivariate calibration problems found in Chapter 5, the concepts are applicable to a much broader range of problems. The following three questions can be used to initiate the discussions.

1. What are the properties of the system to be predicted?
   (These are referred to as the prediction variables, c's, or **C**.)
2. What analytical techniques are being considered to use for measurement?
   (These are referred to as the response variables, r's, or **R**.)
3. What other physical or chemical phenomena affect **C** and/or **R**?

For example, assume the primary objective is to predict the concentration of component A (**C**) given spectroscopic measurements (**R**). This addresses only the first two questions above. There are potentially many other physical or chemical phenomena that can affect **R** and/or **C**; these must also be considered. Additional variables to consider include other components in the samples, the temperature of the sample, and potential impurities. These variables often do not appear in the prediction equation, but nonetheless must be considered when constructing the calibration model because they can have significant impact on the **R** matrix (see Chapter 5 and Appendix A).

From these questions, often employed in a brainstorming session, the necessary information is collected to design the experiments. One way of organizing this information is by using a template, as shown in Table 2.4. At this stage the only objective is to fill in a comprehensive list of variables (column 2). For this hypothetical example, nine variables have been identified in addition to the prediction and measurement variables. The additional variables are both chemical and physical in nature and have been identified as factors that could affect **R** and/or **C**.

Step 2—Characterize the Variables

Once a list of potentially influential variables is identified, the next step is to qualitatively characterize the impact of these variables on **R** and/or **C**. The four properties that are used to accomplish this include: (1) range, (2) controllability, (3) measurability, and (4) complexity. These properties are used to estimate the magnitude of the effect of a variable and help determine possible approaches to account for these effects. For example, the decision might be to set a controllable variable at a fixed value to insure it does not add variability to the resulting data. By characterizing the variables, decisions can be made as to how to handle all of the identified variables based on their specific properties.

The first characteristic considered is the range of the variable. This is simply the maximum and minimum value the variable can take (see column 3 in Table 2.4). Consider the maximum possible range as well as the normal range of variation. A very small range of variation may mean that a variable can be considered to be fixed and thus eliminated from further consideration.

One of the objectives in filling out this template is the elimination of non-significant variables. Referring to the example shown in Table 2.4, assume it is determined that the materials of construction have no impact on the experiment. This variable can therefore be eliminated (note the mark in column 1). Likewise, the analyzer enclosure temperature and isomer impurity level are eliminated because it is determined that they will not influence **R** or **C** over the ranges specified.

There are six variables remaining in addition to the prediction and measurement variables. The next step is to determine whether these variables can be controlled (see column 4). If a variable can be controlled, then it can be either

TABLE 2.4. A Template for Organizing Variable Information when Preparing to Design Experiments

Identify		Characterize			
Eliminate?	Variable Name	Range	Controllable?	Measureable?	Complex (c) Simple (s)
	Component A (**C**)	50–70%	N	Y	s
	Spectroscopy (**R**)	—	—	Y	—
	Water	0–10%	N	N	c
X	Isomer impurity	0.010–0.012%	—	—	—
X	Sample temperature	20–25°C	Y	Y	c
X	Instrument enclosure temperature	22.5–23°C	—	—	—
	Pressure	1–2 atm	N	Y	—
	Oxygen concentration	0–2000 ppm	N	N	—
	Analyte Z	5–8%	N	Y	s
	Analyte B	20–40%	Y	Y	s
X	Material of construction	—	—	—	—

fixed or varied during the experiments. The easiest way of dealing with a variable is to fix it at a constant level. By doing this, it is essentially removed from the analysis in that it will not add variability to the data. This means the variable *must* remain fixed in the future for the method to work correctly. If fixing the variable is not possible, the effect of the variable can be accounted for in the model by varying the levels during data collection.

The next column (column 5) in the template indicates whether the magnitude of the variable can be measured during experimentation. In cases where a variable is controllable, it is arguable that it also must be measurable. However, the converse is not necessarily true, that is, a variable that cannot easily be controlled often can be measured (e.g., pressure in a reactor). Measuring a variable, even when it cannot be controlled, is useful for two reasons: (1) the range of a variable can be tracked during experimentation and (2) the measured values can be used in an explicit model.

The last column in the template is labeled "Complex (c); Simple (s)." This indicates the complexity of the effect the variable is expected to have on **C** and/or **R**. This is another important characteristic to consider because it directly affects the experimental design. In general, the complexity of the variable is considered simple, "s," when it is expected to have a linear effect. All other relationships are considered to be complex and are indicated by a "c" in the template. This knowledge may come from experience with a particular system, physics, or other sources of information. The complexity is used in the design section to determine how many levels of the variables are needed in the design.

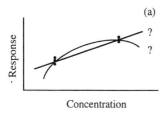 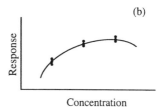

Figure 2.3. Two experimental designs for modeling a potentially nonlinear system. (a) Two point calibration does not detect curvature; (b) three point calibration can account for quadratic behavior.

An examination of the two designs found in Figure 2.3 illustrates why this information is important. Designs (a) and (b) have two and three levels of concentration and a corresponding hypothetical response curve. Design (a) can only model a linear relationship; two points determine a straight line. With only two levels, there is not enough information to test for the existence of a nonlinear relationship. With design (b) it is possible to test for and model a quadratic relationship. This is a very simple example, but the same idea translates to more complex multivariate models. The point is that more levels are needed in the design to model more complex relationships.

In Table 2.4 analytes Z and B are considered simple, while water concentration and sample temperature are complex. This assumes that the interfering analytes Z and B approximately obey Beer's Law and that there will be no interaction effects. Water is considered complex because there is a possibility that the detector will become nonlinear at higher concentrations. The temperature was considered complex because of band broadening and potential effects on the viscosity of the sample.

Step 3—Specify the Design
At this point an attempt has been made to identify all of the important variables. It has also been decided which variables will be fixed and which will be varied. For the nonfixed variables, the range and number of levels (complexity) have been determined. Classical experimental design tools will be used to specify the design for the controllable variables. The design is selected via a software package or statistical reference materials and it specifies: (1) the number of runs, (2) the levels of the variable(s), and (3) the order of the runs.

Although it is recommended that classical designs be used whenever possible, it is common to deviate from the standard statistical designs. Either it is cost prohibitive to run the required number of experiments, design points are not chemically or practically feasible, or some other factor precludes the exact use of the design. (See Appendix A for special considerations when designing experiments for inverse chemometric models.)

The design that is chosen also depends on the stage of the research and the resources available. In early stages the goal is often to confirm the importance of the variables that have been identified. The screening designs found in

TABLE 2.5. Stages of Research and Types of Classical Experimental Designs

Type of Experiment	Number of Variables	Typical Information Sought	Typical Design
Screening	6–40	Identify important variables.	Plackett–Burman (PB)
		Estimate approximate effects.	2^n Fractional factorial (FF)
Linear effects	3–8	Prediction over limited region. Estimate linear effects and interactions only.	2^n Full and fractional factorial, PB with foldover
Response surface	2–6	High-quality prediction for linear, interaction, and curvature.	Central composite, Box–Behnken, 3^n factorial
Mechanistic model	1–5	Accurate prediction inside and outside the range of calibration data.	Special designs (usually iterative)

Table 2.5 have relatively few design points and are efficient designs to use for this purpose. These designs are often used in early stage research. If more information is sought or the relationship is more complex, the linear effects designs might be more appropriate. Response surface designs are considered if even more detailed information is required and/or the relationship to be modeled requires quadratic terms. If the need is for very accurate prediction and extrapolation, special designs can be used to develop mechanistic models.

The use of these classical experimental designs requires that the variables be set to predetermined levels. Therefore, additional effort must be made to account for variables that are not controllable. One chemometric approach is to allow these variables to vary naturally and to collect enough data to adequately model their effect. This is the difference between the so-called natural and controlled calibration experiments (Martens and Næs, 1989). When it is possible to measure the variables, this can be done to verify that an adequate range has been covered. Inverse models as discussed in Chapter 5 can then be used to implicitly model their effect. (See also Appendix A.)

Recall that in the example above the interest is in developing a predictive model for component A using spectroscopy. A response surface design is appropriate for the controllable variables because the model is to be used for prediction and the relationship of some of the variables is considered to be complex. Table 2.4 also shows that the pressure and oxygen concentration cannot be controlled, but the variation is significant. In this case, a natural design for these two variables also needs to be incorporated into the experimental scheme. An inverse calibration technique can then be used to develop a predictive model.

2.4   SUMMARY OF DEFINING THE PROBLEM

Understanding the system and the problem, and effectively communicating possible solutions to the team, are some of the most important steps for a successful project. Asking pertinent questions, listening, learning, and teaching are all part of the process of defining the problem. Using the entire team's knowledge base to define the problem helps insure that the right problem is specified. Once this is accomplished, the variables are characterized and the experimental design can be selected using statistical design software.

REFERENCES

G. E. P. Box, W. Hunter, and J. S. Hunter, *Statistics for Experimenters,* Wiley, New York, 1978.

S. N. Deming and S. L. Morgan, *Experimental Design: A Chemometric Approach,* Elsevier, New York, 1987.

N. Draper and H. Smith, *Applied Regression Analysis,* 2nd ed., Wiley, New York, 1981.

H. Martens and T. Næs, *Multivariate Calibration,* Wiley, New York, 1989.

APPENDIX A.   SPECIAL DESIGN CONSIDERATIONS FOR CHEMOMETRIC INVERSE MODELING

Experimental planning concepts are emphasized in Chapter 2 because it is very important to carefully consider the variables that affect the data before they are collected. The concepts taught in Chapter 2 are most appropriate for classical regression modeling. When using inverse modeling methods in chemometrics, a modification to the classical approach is required (see Section 5.3 for details on inverse models). To explain why this is necessary, the standard regression model shown in Equation A.1 is discussed.

$$\mathbf{y} = \mathbf{Xb} \qquad (A.1)$$

In Equation A.1, the vector $\mathbf{y}$ (nsamp x 1) is called the dependent variable and contains one response for each of the nsamp samples. There are nvars independent variables found in the columns of the matrix $\mathbf{X}$ (nsamp x nvars). Each sample in the data set is represented by a row in this matrix, and $\mathbf{b}$ is the vector of regression coefficients estimated using regression analysis. The columns of the $\mathbf{X}$ matrix are the variables that are identified and defined using the process found in Chapter 2. In the classical approaches to modeling, the goal of the regression analysis is to understand the effect the columns of $\mathbf{X}$ have on the $\mathbf{y}$ variable. For example, the columns of $\mathbf{X}$ might be temperature, pressure, and humidity and the interest is in how these affect the permeability of a membrane ($\mathbf{y}$). To determine the effects, a set of experiments is designed whereby the independent variables ($\mathbf{X}$) are set at different levels and the dependent variable ($\mathbf{y}$) is measured. The combinations of settings used to generate $\mathbf{X}$ defines the experimental design. This is the type of problem for which the classical experimental designs are best suited.

To see how the inverse and classical modeling approaches differ, the inverse model represented by Equation A.2 is examined.

$$\mathbf{c} = \mathbf{Rb} \qquad \qquad (A.2)$$

At first glance this equation does not look any different than the classical model in Equation A.1. In fact, the goals of these two models are similar, but the processes of collecting data and estimating the models are different. The letters $\mathbf{c}$ and $\mathbf{R}$ are used in Equation A.2 to represent sample characteristics or concentrations ($\mathbf{c}$) and measurement responses ($\mathbf{R}$). The columns of $\mathbf{R}$ can be, for example, spectral measurements at different wavelengths where each wavelength is a measurement variable. The goal in this analysis is similar to the classical approaches in that regression coefficients ($\mathbf{b}$) are to be estimated. With the inverse model the ultimate goal is to use these coefficients to predict the concentrations ($\mathbf{c}$) from the measurements ($\mathbf{R}$) made on unknown samples.

If a classical approach were used to estimate the regression coefficients, the process would be to set the variables in the matrix $\mathbf{R}$ to predetermined values as dictated by the experimental design. This does not make sense when the measurement system for $\mathbf{R}$ is spectroscopy. One cannot choose to set the different wavelengths to fixed values and collect concentration information on the corresponding samples. What can be controlled is the concentration of the components in the samples (i.e., $\mathbf{c}$). The approach, therefore, is to choose samples with varying concentrations and measure the spectra on these samples. This is the opposite of the classical approach where the independent variable ($\mathbf{X}$) is set and the dependent variable ($\mathbf{y}$) is measured.

Fortunately, with well-behaved systems, classical statistical designs can still be employed when using the inverse approach. These design techniques can be used to design the concentration vector or matrix as long as all sources of relevant variation that can affect the responses are taken into account. This introduces another concept that is important to understand. With the classical approach, it is easy to envision how variables are taken into account. If temperature has an effect on the dependent variable, it is either measured and incorporated into the model as a column in $\mathbf{X}$ or it is held constant. This is not as straightforward for an inverse model. Returning to the spectroscopic example, if the goal is to estimate a concentration with spectral data, how can the effect of varying temperature on the spectroscopy be taken into account? An inverse modeling technique can be used to implicitly incorporate the effects of the variable into the prediction model (see inverse methods in Chapter 5).

Implicit modeling means that variation is captured without explicit identification of the variable causing the variation. All that is required is that sufficient variation in the level of the variable be present in the samples. Using this approach, it is not necessary to explicitly measure the additional variable, that is, its level does not have to be recorded. This is because the variation is captured via the **R** matrix.

An example of inverse modeling is given in the PLS example in Section 5.3.2.2. In this example, the concentration of caustic in aqueous solutions is estimated even when the salt content and temperature are varying. Measurements of the temperature or salt concentration are not used during the model-building phase, even though they have a significant effect on the spectroscopy. By collecting enough spectra at varying salt concentrations and temperatures, PLS is able to distinguish the spectral variations associated with the caustic from those associated with these other effects.

In this particular example, the salt and temperature information was recorded, however with the inverse modeling approach, the values are not used in the computation of the PLS model. One might be tempted to want to account for these variables by including them as additional columns of **R**. However, this is not necessary, because the effects of these variables are already captured by the spectra. Complimenting the **R** matrix with variables related to or correlated with the **c** vector may be helpful if that correlation is different from what is already in **R**. This is in contrast to a more classical approach for analyzing these same data, discussed in Section 5.2.2.2.

One last point to discuss is the complexity or type of experimental design to use. In the classical case, the appropriate type of design depends on the complexity of the postulated model. When using inverse methods, the recommendation is to assume at least a quadratic relationship. This means three or more levels of concentrations are required for the design. The class of designs that fulfill this requirement are termed response surface designs (see Table 2.5). Potential calibration designs include the full factorial design with three levels or fractional factorial designs if the number of analytes to be modeled is large. Other designs may be considered if more nonlinear behavior is expected. A central composite design (CCD) is also a good choice for many problems. (See Chapter 2 for references on experimental design.)

Another approach for experimental design for inverse modeling is termed a natural design. This is where many samples are collected over a period of time until one has confidence that the variation has been adequately represented. In some cases, these natural designs are the only choice because neither **R** nor **c** can be controlled. Successful use of this approach requires some knowledge of the system in order to make an intelligent assessment of how many samples are required and how long to sample. One rule of thumb is to have at least three times as many samples as the expected rank of the system. (Rank is a concept that is discussed in Chapters 4 and 5.)

CHAPTER

3

Preprocessing

Be prepared.

—Motto of The Boy Scouts of America

Preprocessing is a very important part of any chemometrics data analysis project. It is so important that it is delineated as one of the "Six Habits of an Effective Chemometrician" (see Chapter 1) and is defined as any mathematical manipulation of the data prior to the primary analysis. It is used to remove or reduce irrelevant sources of variation (either random or systematic) for which the primary modeling tool may not account. Keep in mind that preprocessing changes the data which will either positively or negatively influence the results. "Being prepared" by applying the appropriate preprocessing tool(s) is critical in order for the overall data analysis to be successful.

Selecting the optimal preprocessing may require some iteration between the primary analysis and the preprocessing step. Although this empirical approach is a common practice, it is best if the preprocessing tool is chosen because of a known characteristic of the data. For example, percent transmission spectra are often linearized with respect to concentration by converting them to absorbance units.

In this chapter a number of preprocessing tools are discussed. They are divided into two basic types depending on whether they operate on samples or variables. Sample preprocessing tools operate on one sample at a time over all variables. Variable preprocessing tools operate on one variable at a time over all samples. Therefore, if a sample is deleted from a data set, variable preprocessing calculations must be repeated, while the sample preprocessing calculations will not be affected.

3.1 PREPROCESSING THE SAMPLES

The first set of preprocessing tools discussed are those that operate on each sample. Table 3.1 lists the four methods discussed: normalizing, weighting, smoothing, and baseline corrections. Normalization can be used to remove

TABLE 3.1. Sample Preprocessing Tools

Method	Use
Normalizing	Puts all the samples on the same scale by dividing by a constant (e.g., removing variable injection volume in chromatography).
Weighting	Sample weighting gives some samples more influence on the analysis than others (e.g., a weight of zero eliminates a sample).
Smoothing	Reduces the amount of random variation (noise).
Baseline corrections	Reduces systematic variation.

sample to sample absolute variability (e.g., variable injection volumes in chromatography) while weighting emphasizes selected samples over others. Smoothing is primarily used to reduce random noise whereas the other sample preprocessing methods are used to remove systematic variations. Baseline features can be removed using explicit models, derivatives, or multiplicative scatter correction.

3.1.1 Normalization

Normalization of a sample vector is accomplished by dividing each variable by a constant. Different constants can be used and three are described here.

Normalizing to unit area is accomplished by dividing each element in the vector by the "1-norm." The 1-norm of a vector is the sum of the absolute value of all of the j entries in the vector $\mathbf{x}$, as shown in Equation 3.1.

$$\text{1-norm} = \sum_{j=1}^{\text{nvars}} |x_j| \qquad (3.1)$$

Normalizing to unit length is accomplished by dividing each element in the vector by the "2-norm." The 2-norm is calculated by taking the square root of the sum of all the squared values in the vector, as shown in Equation 3.2.

$$\text{2-norm} = \sqrt{\sum_{j=1}^{\text{nvars}} x_j^2} \qquad (3.2)$$

Normalizing so that the maximum intensity is equal to 1 is accomplished by dividing each element in the vector by the infinity norm, defined as the maximum (in absolute value) of the vector.

Normalization is performed in order to remove systematic variation, usually associated with the total amount of sample. A common example of this is normalizing to the largest m/e peak in mass spectrometry (Howe et al., 1981, p. 19). In chromatography, normalization of the entire chromatogram to unit area is used to remove the effect of variable injection volume. Normalizing to unit area is also used in library searching in mass spectrometry (Howe et al., 1981, p. 229), and in two-component curve resolution. (Lawton and Sylvestre, 1971).

Consider an example of a chromatographic application with two components of interest. The raw chromatograms from two injections of two samples are shown in Figure 3.1a. The chromatograms normalized to unit area (1-norm) displayed in Figure 3.1b demonstrate the elimination of injection volume variations (i.e., the chromatograms from the same sample overlay).

A second example of normalization comes from an application of near-infrared reflectance spectroscopy for sorting recycled plastic containers. Spectra of discarded containers were measured and pattern-recognition tools were applied in order to facilitate the sorting (see Sections 4.2.1.2 and 4.3.1.2). Prior to applying the pattern-recognition tools, extensive preprocessing was performed. Shown in Figure 3.2a are spectra of the polyethylene samples (the second derivative of the data has already been taken for reasons discussed later). Due to variations in the pathlength, the spectra vary in intensity. Normalization to unit area (dividing by the 1-norm) reduces this pathlength variation (see Fig. 3.2b). The spectra overlay more closely, especially in the 1750-

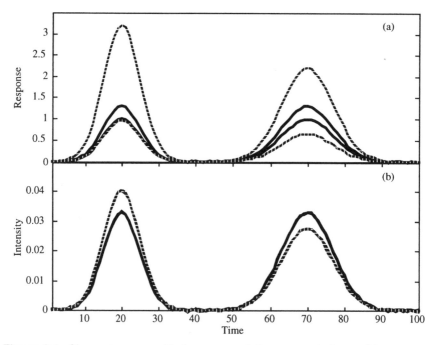

Figure 3.1. Chromatograms with the same relative concentrations of two components but with varying injection volumes. Results before (a) and after (b) normalization to unit area.

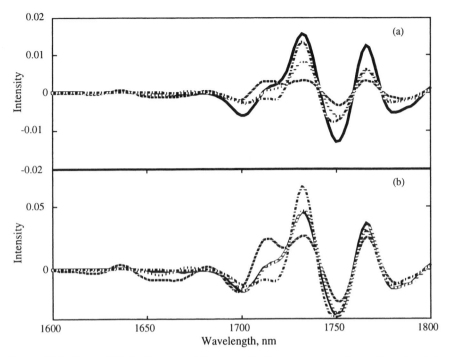

Figure 3.2. Second derivative near-infrared reflectance spectra of recycle polyethylene containers before (*a*) and after (*b*) normalization to unit area.

to 1800-nm region. The remaining spectral differences are due to chemical variations in the samples.

In summary, normalization reduces systematic variation in the data by dividing by a constant. Depending on what variation is to be removed, one normalization constant may be more appropriate than another. Beware that normalization may remove important concentration information.

3.1.2   Sample Weighting

Sample weighting is accomplished by multiplying each element in a sample vector by a constant. In this way, the influence a sample has on a mathematical model can be manipulated. Sample weighting is similar to normalization, but the criteria for defining the constants differ. The weights can be any values, although weighting should only be applied when reliable information is available about the relative importance of some samples over others. For example, the data from a highly experienced analyst can be given more weight than those of a trainee. Another use of weighting is to satisfy inherent assumptions of the primary method of analysis (e.g., the assumption of homoscedastic errors in linear regression, Draper and Smith, 1981).

3.1.3 Smoothing

In analytical chemistry, it is assumed that a measured signal consists of the true signal plus random noise. The amount and structure of the noise depends on the experiment. Smoothing tools (smoothers) are used to mathematically reduce the random noise with the goal of increasing the signal-to-noise ratio. A basic assumption made with these tools is that the noise is of higher frequency relative to the signal of interest. It is the redundant information contained in adjacent variables that enables smoothers to separate the "true" signal from the noise. Some sources of noise contribute to the low-frequency signal, but they are often difficult to mathematically remove without removing some of the chemical information of interest. Baseline correction methods are used to remove these low-frequency signals.

Smoothing methods typically use a window which can be thought of as a region of influence. All the points in the window are used to determine the value at the center of the window, and therefore the window width directly affects the resulting smooth. Five methods for smoothing are discussed below. Four of them use a window, but differ in how the points in the window "vote." The fifth method, Fourier smoothing, does not use a window.

3.1.3.1 MEAN SMOOTHER As defined here, a mean smoother is used to decrease the number of variables in a sample vector. This may be needed if, for example, the calculation speed must be increased. To begin, a window width (n) is chosen and the mean of the first n points in the sample vector is calculated. This defines the first entry in the mean smoothed vector. The second entry is calculated as the mean of the $n + 1$ to $2n$ points in the original sample vector. This process is repeated for all elements in the original vector. The resulting smoothed vector has a factor of n fewer elements. The mean smoother with a reasonable window width is better than contracting the vector by taking every nth point because the mean calculation results in signal averaging. Figure 3.3a displays a spectrum containing 800 variables. In Figure 3.3b, a mean smoother with a window width of 20 has been applied which has reduced the number of variables to 39. The mean smoother always reduces the resolution; this is evidenced in Figure 3.3b by the elimination of the sharper features. Therefore, choosing an inappropriate window width may eliminate important features in the data.

3.1.3.2 RUNNING MEAN SMOOTHER Running smoothers operate by moving the window across the data vector one element rather than one window width at a time as with the mean smoother described above. This results in a smoothed vector that is the same (or almost the same) length as the original sample vector. Specifically for the running mean smoother, the jth element in the new vector is the mean of the original data located in the window centered around the jth element. These smoothers introduce features in the ends of the sample vector and are termed "end effects." For example, the first element in the vector is often deleted because it cannot be in the middle of a window. In fact (window size-1)/2 points on either end of the vector cannot be smoothed in the same manner as the remaining points.

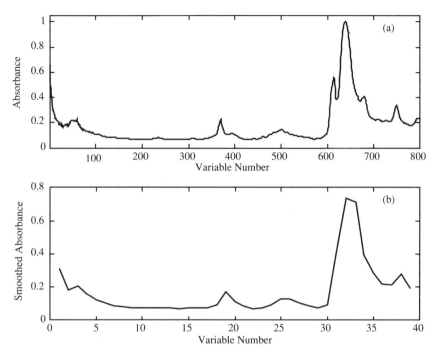

Figure 3.3. Spectrum before (*a*) and after (*b*) applying a mean smoother with a window width of 20.

This smoother is one way to improve the signal-to-noise ratio of the data, as shown in Figure 3.4. The original data has some visible random noise and a large spike. The mean smoother with a 3-point window reduces the noise significantly, but does not remove the spike. With the largest window size (21 point), the spike is removed but the shape of the peak has changed (broader and lower intensity). The apparent shift in the peak to lower variable number is due to the end effect (i.e., 10 points from each end of the sample vector have been removed).

3.1.3.3 RUNNING MEDIAN SMOOTHER

The running median smoother is similar to the running mean smoother except the median is used instead of the mean. The median is not as sensitive to extreme points as the mean (Hoaglin et al., 1983) and, therefore, the median smoother is very effective at removing spikes from the data. However, it is not as efficient at filtering noise. The median smoother applied to the raw data presented in Figure 3.4 is shown in Figure 3.5. Compare the 3-point window results in Figures 3.4 and 3.5. The median smoother removed the spike better than the mean smoother, but the latter more effectively reduced the noise. Because of the complementary nature of the two approaches, a combination of running mean and running median smoothers may be appropriate for some data sets.

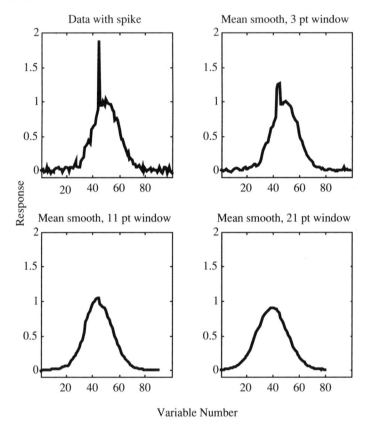

Figure 3.4. Results after applying a running mean smoother with varying window widths to a sample vector.

3.1.3.4 RUNNING POLYNOMIAL SMOOTHER The running polynomial smoother differs from the running mean and median smoothers in that a low-order polynomial is fit to the points in the window. The jth element in the smoothed data vector is equal to the polynomial prediction at element j. A convenient implementation of this approach is that of Savitzky and Golay (1964). An example of a polynomial fit over one window width (13 points) is shown in Figure 3.6. The data in this window are used to calculate the smoothed value in the middle of the window. The solid line shows the second-order polynomial fit to the data, and the "X" is the smoothed value for data point 7.

Using the Savistky–Golay method results in the elimination of (window size-1)/2 points on each end of the sample vector. If this is unacceptable, Gorry has developed a method that does not result in the elimination of points (Gorry, 1990, 1991). While this method preserves the original number of variables, it can introduce aberrant features to the ends of the sample vectors (Hui and Gratzl, 1996). Overall, the Gorry method is recommended and is used in all subsequent discussions.

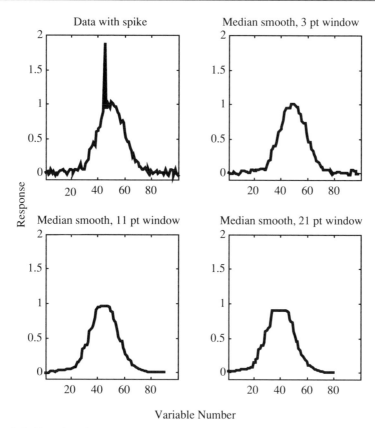

Figure 3.5. Results after applying a running median smoother with varying window widths to the same data found in Figure 3.4.

As with all window-based smoothers, the choice of the window size in polynomial smoothers is very important. Another decision to make for polynomial smoothers is the order of the polynomial to be fit (Barak, 1995). Typically, a second- or third-order polynomial is used. An example of applying a polynomial smoother is shown in Figure 3.7, where a second-order polynomial is fit with window sizes of 7, 13, and 25 points. As the window size increases, the noise is continually reduced. However, when the window is too large, sharp peaks may be removed and the remaining peaks distorted. This is demonstrated in Figure 3.8 where a spectrum is shown before (solid) and after (dashed) applying a 49-point second-order polynomial smoother.

For a given data set, the optimal window size and polynomial order depend on the nature of the data. Of primary importance is the width of the peaks relative to the window width (e.g., choosing a window width 10 times the width of a peak will most likely distort or eliminate it). An approach to selecting a reasonable window width and polynomial order is to apply several combinations and evaluate the resulting preprocessed data and final results.

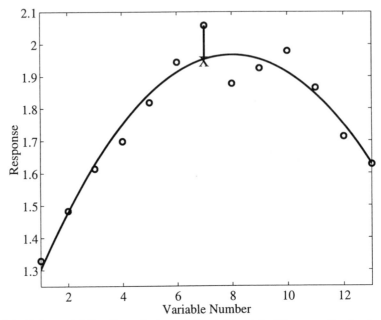

Figure 3.6. Polynomial fit with a window width of 13 points. The smoothed value of data point 7 is shown as X.

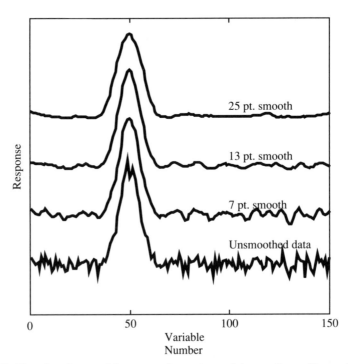

Figure 3.7. Results after applying a running polynomial smoother with varying window widths to a sample vector. (The offset was added for clarity.)

34

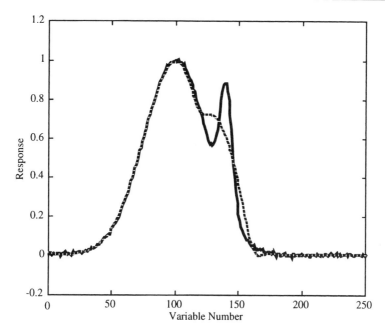

Figure 3.8. Illustration of peak distortion when using a smoother with a too-large window width (49 point). (Solid—raw data; dashed—smoothed data.)

3.1.3.5 FOURIER FILTER SMOOTHER Fourier filtering can be used for general smoothing or removal of specific frequency components. Both of these are accomplished by Fourier transforming the signal, weighting the transform with an apodization function, and back transforming to the original units.

With so many fields using Fourier analysis, the notation is varied and sometimes conflicting (Table 3.2 lists some of the notation). Interestingly, the infrared spectroscopy and mathematics notation are completely opposite. Because of this ambiguity, we will define column 3 in Table 3.2 as the apodization domain because this is where the apodization is always applied. For lack of a better term, we will refer to column 2 as the time domain.

General Fourier filter smoothing is accomplished by using an apodization function in the apodization domain. The interferogram is multiplied by the apodization function before transforming to the time domain. There are many types of apodization functions (Griffiths and de Haseth, 1986), perhaps the most simple being boxcar apodization. One example of boxcar apodization is zeroing high-frequency Fourier coefficients. Figure 3.9*a* displays a data vector that is transformed to the apodization domain in Figure 3.9*b*. The first and last 50 points of this interferogram are set to zero, as displayed in Figure 3.9*c*. This is transformed back to the original units, as shown in Figure 3.9*d*. This final sample vector is more smooth than the original vector because the high-frequency components are removed. Changing the number of Fourier coefficients that are zeroed in Figure 3.9*b* yields a different result. Figures 3.10*a* and

TABLE 3.2. Fourier Analysis Notation

	Time Domain	Apodization Domain
Typical Graph		
Chromatography	Time	Frequency
FT–NMR	Shift	Relaxation time FID
FTIR	Frequency (e.g., wavelength,	Position, time, interferogram
FTNIR	wavenumber)	
Mathematics	Time	Frequency

b display the results of zeroing 70 coefficients rather than 50. This has produced a significantly broadened peak and has introduced artifacts to the baseline.

Fourier filtering can also be used to remove a specific frequency present in the data. Common examples of this include removing baseline offsets (low frequency) and 60 Hz line noise. Figure 3.11*a* displays a sample vector with a periodic feature. After the Fourier transform has been applied, this feature appears in the interferogram as a single point at variable 100 (Figure 3.11*b*). Figure 3.11*c* has this frequency zeroed, which when back transformed results in smoothed data without the periodic noise (Figure 3.11*d*).

This concludes the discussion of smoothers; a summary with recommendations is provided in Table 3.3.

3.1.4 Baseline Corrections

Besides high-frequency components (noise), the measured signal may also contain low-frequency sources of variation that are not related to the chemistry under investigation. In this book, these components are called baseline features. These systematic variations can be large relative to changes in the signal of interest and may dominate the analysis if not removed. They may also vary randomly in intensity and shape from sample to sample. A number of approaches for reducing baseline features are discussed below.

3.1.4.1 EXPLICIT MODELING APPROACH Any sample vector can be written as a function of variable number ($\mathbf{x}$),

$$\mathbf{r} = f(\mathbf{x}) \tag{3.3}$$

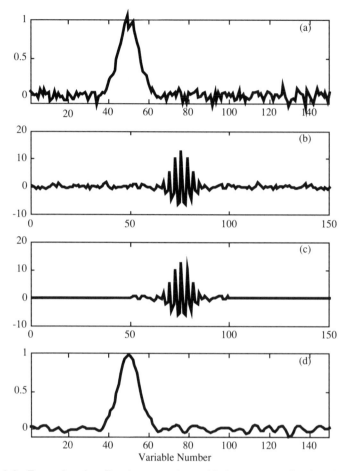

Figure 3.9. Example of a Fourier smoother with boxcar apodization (zeroing 50 points on each end). (a) Original spectrum; (b) spectrum in apodization domain; (c) first and last 50 points set to zero; (d) after transformation back to original units.

This function is equal to the sum of the signal of interest plus some baseline feature (if present). The baseline can be approximated using a polynomial, as shown in Equation 3.4

$$\mathbf{r} = \tilde{\mathbf{r}} + \alpha + \beta\mathbf{x} + \gamma\mathbf{x}^2 + \delta\mathbf{x}^3 + \ \ldots \tag{3.4}$$

where $\tilde{\mathbf{r}}$ is the signal of interest and the remainder of the equation approximates the baseline feature. By postulating a model for the baseline (e.g., offset, linear, polynomial), one can account for it directly by subtraction. For example, a sample vector with an offset baseline feature (i.e., a horizontal line) can be written as

$$\mathbf{r} = \tilde{\mathbf{r}} + \alpha \tag{3.5}$$

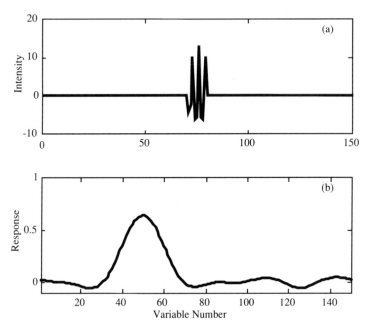

Figure 3.10. Example of a Fourier smoother with boxcar apodization (zeroing 70 points on each end). (a) Spectrum in Figure 3.9 in apodization domain with first and last 70 points set to zero; (b) after transformation back to original units.

In this case, the baseline can be removed by estimating α and subtracting it from the sample vector ($\mathbf{r}$). This is illustrated in Figure 3.12a, where a series of sample vectors with offset baseline features is plotted. The offset can be removed by subtracting the intensity of a variable from all variables for each sample vector. The optimal variable is one that contains only baseline information (this is α in Equation 3.5). In this example, variable 60 is used to estimate α, which is subtracted from all elements in the sample vector. The preprocessed data shown in Figure 3.12b now reveal two groups of samples which were not apparent prior to preprocessing. Note also in Figure 3.12b that all sample vectors have zero intensity at variable 60. The average intensity of several baseline variables is often used to estimate α. This yields a better estimate of α and reduces the amount of noise introduced into the sample vectors by the baseline subtraction.

Another example of explicit baseline modeling is presented in Figure 3.13, where the sample vector contains a linearly sloping baseline. This type of baseline is encountered in chromatography as well as spectroscopy. It can be due, for example, to solvent gradients in chromatography or wavelength-dependent scattering in spectroscopy. Mathematically, this can be represented as

$$\mathbf{r} = \tilde{\mathbf{r}} + \alpha + \beta \mathbf{x} \tag{3.6}$$

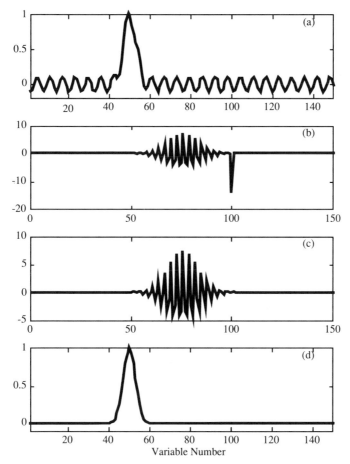

Figure 3.11. Example of eliminating a periodic noise component in a sample vector using a Fourier smoother. (a) Original spectrum with 60 Hz line noise; (b) spectrum in apodization domain; (c) spike at variable 100 set to zero; (d) after transformation back to original units.

To remove the baseline component, a line is estimated (α and β) using two or more points that are assumed to contain only baseline information. The estimated line for this example is shown as a dashed line in Figure 3.13a. To remove the baseline feature, this line is subtracted from the sample vector as shown in Figure 3.13b.

Other functions can be estimated if the baseline has a more complex shape. Regardless of the function used, the key is to choose points for estimating the coefficients in Equation 3.4 (α, β, γ, . . .) that are only influenced by the baseline. If the points are chosen poorly, a portion of the chemical variation will be removed in addition to the baseline.

TABLE 3.3. Summary of Smoothing Tools

Method	Summary and Recommendations
Mean	Use this method to reduce the number of variables, but beware of the consequences of lower resolution. It is preferred over simply taking every nth data point because of the signal averaging that results from calculating the averages.
Running mean	This method works reasonably well for general smoothing. Use if no better smoothing methods are available.
Running median	Use for removal of high-frequency spikes. Not as efficient as the running mean smoother for noise reduction.
Running polynomial	Preferred method for noise reduction. The method of Gorry is recommended because no truncation of data occurs.
Fourier filtering	Good method for general smoothing, but must select an appropriate apodization function. Best method for removing specific periodic features in the raw data provided the corresponding frequency(ies) can be identified in the apodization domain.

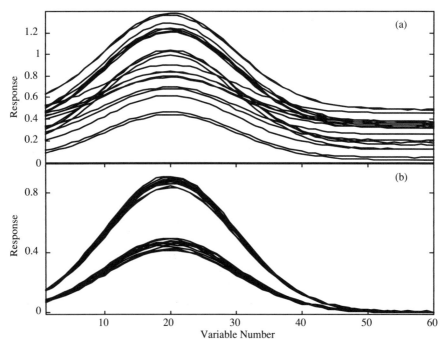

Figure 3.12. Data with a baseline offset before (*a*) and after (*b*) baseline correction using the explicit modeling approach.

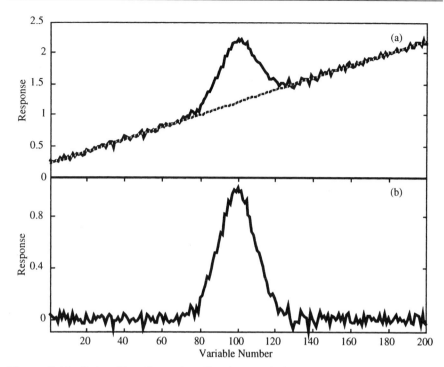

Figure 3.13. Data with a linear baseline feature before (*a*) and after (*b*) baseline correction using the explicit modeling approach.

3.1.4.2 DERIVATIVES Another way of removing baseline features is to take derivatives with respect to variable number. By using this approach, it is not necessary to select points that only contain baseline information. This is especially useful in cases where baseline points are difficult or impossible to identify.

To see how the derivatives are able to remove baseline features, refer back to Equation 3.4. Taking the first derivative of the sample vector with respect to variable number ($\mathbf{x}$) yields $\mathbf{r}'$,

$$\mathbf{r}' = \tilde{\mathbf{r}}' + 0 + \beta + 2\gamma\mathbf{x} + 3\delta\mathbf{x}^2 + \ldots \tag{3.7}$$

where $\tilde{\mathbf{r}}'$ is the derivative of the signal of interest. Equation 3.7 reveals that the first derivative has completely removed the offset feature (α). If the baseline is only comprised of an offset, the other coefficients in Equation 3.4 (and therefore Equation 3.7) would be zero.

If a more complex baseline is present (β, γ, δ, $\ldots \neq 0$), repeated application of the derivative will successively remove the higher-order terms. For example, taking the derivative of Equation 3.7 yields

$$\mathbf{r}'' = \tilde{\mathbf{r}}'' + 0 + 0 + 2\gamma + 6\delta\mathbf{x} + \ldots \tag{3.8}$$

Equation 3.8 is equivalent to the second derivative of Equation 3.4 and the linear feature (α and β) has been completely removed.

Running Simple Difference

The simple difference between adjacent data points can be used to estimate the first derivative. For a sample vector $\mathbf{r} = [r_1, r_2, \ldots, r_n]$ the first derivative can be estimated as $\mathbf{r}' = [r_2 - r_1, r_3 - r_2, \ldots, r_n - r_{(n-1)}]$. This procedure can be repeated to estimate the second and successive derivatives. As an example, Figure 3.14a displays a noise-free Gaussian peak with a constant offset of one unit and Figure 3.14b shows the derivative calculated by simple difference. The baseline has been removed (the region where there was no peak in the raw data now has zero intensity), and as expected, the peak has the shape of the derivative of a Gaussian.

When noise is present, the simple difference approach for calculating a derivative is not effective. The difference calculation propagates errors into the derivative which degrades the signal-to-noise. This is illustrated in Figures 3.15a and b where the same Gaussian peak with noise added and the simple difference result are shown, respectively. It is clear that the signal-to-noise has been decreased by the preprocessing.

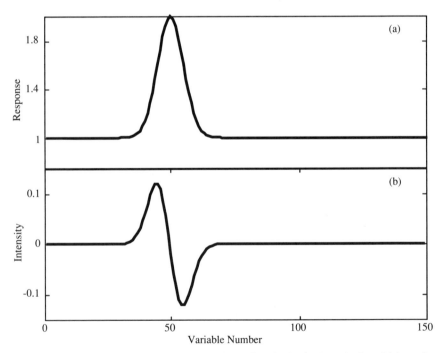

Figure 3.14. Noise-free data with a baseline offset before (*a*) and after (*b*) baseline correction using a simple difference derivative.

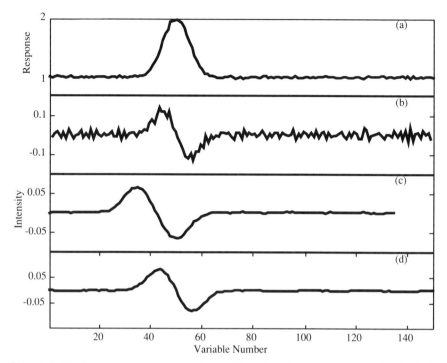

Figure 3.15. Results after applying three different derivative preprocessing tools to a sample vector. (*a*) A sample vector with noise and an offset of one unit. (*b*) The derivative calculated by simple difference. (*c*) The derivative calculated using a running mean difference with a window width of 15. (*d*) The derivative calculated using the Gorry method with a window width of 15.

Running Mean Difference

One way of improving the error-propagation properties of the simple difference method is to use the difference of means for the derivative calculation. With the running mean difference, a window width is selected and the difference between means is taken instead of using the difference between individual points. For example, using a window size of 3, the first derivative of the sample vector $\mathbf{r} = [r_1, r_2, \ldots, r_n]$ can be estimated as

$$\mathbf{r}' = \left[\frac{r_2 + r_3 + r_4}{3} - \frac{r_1 + r_2 + r_3}{3}, \ldots \right] \tag{3.9}$$

This results in a smoothed derivative calculation (see Section 3.1.3 for discussion of smoothing). Figure 3.15*c* shows the result of applying a running mean difference derivative with a window width of 15 to the data in Figure 3.15*a*. The signal-to-noise of this derivative is much better than the simple difference derivative. (The apparent shift of the derivative is due to the end effects.)

The Methods of Savitzky–Golay and Gorry

Another approach to calculating derivatives is based on the Savitzky–Golay and Gorry smoothing methods (see Section 3.1.3.4; Savitsky and Golay, 1964; and Gorry, 1990, 1991). Recall that these methods fit a simple polynomial to a running local region of the sample vector. A window width is selected and the point in the center of the window is replaced with the polynomial estimate of that point. With the derivative, this point is instead replaced with the *derivative* of the polynomial at that point. Because polynomials are used, it is a simple mathematical step to determine the derivative. As with smoothing, the contribution from Gorry was developing a method for treating the ends of the vector so as not to eliminate points. The improvement in signal-to-noise achieved using the method of Gorry with a window width of 15 is demonstrated in Figure 3.15*d* for the first derivative of the data in Figure 3.15*a*. While the signal-to-noise of the running mean and the Gorry methods are similar, we recommend using the Gorry method because fitting a polynomial preserves the peak shape better than the running mean.

To show the importance of smoothing, especially when calculating higher-order derivatives, Figures 3.16*a–c* show a sample vector with a linear baseline, a simple difference second derivative, and a second derivative by the Gorry method using an 11-point window, respectively. Using the simple difference method, propagation of error is more problematic when calculating the second derivative compared to the first derivative (compare Figures 3.15*b* and 3.16*b*). Therefore, using a smoothed derivative method, such as Gorry's, is even more important for this and higher-order derivatives. (In Figure 3.16*c* the Gorry method has introduced aberrant features to the ends of the second derivative. See also Section 3.1.3.4.)

A critical consideration when taking derivatives is the window width for the polynomial fit. If the window size is too small, too little smoothing takes place, resulting in derivatives with poor signal-to-noise. If the window size is too large, features will be smoothed out. The optimal window size depends on the data, because smoothing away features may or may not be detrimental to the primary analysis. The noise level, the number of data points, and the sharpness of the features should all be considered when selecting a window width. The sample vector shown in Figure 3.17 is used to demonstrate the effect of differing window widths. The first derivative results are shown in Figure 3.18. Using a window width of three results in a derivative with poorer signal-to-noise than the original data. As the window width is increased to 21, the signal-to-noise improves, but the peak at variable 55 has been smoothed away. The effect of the window width on the second derivative results in Figure 3.19 is even more dramatic. It is not possible to make a general statement as to which of the window widths is best. In practice, the primary analysis is repeated with several window widths to determine which yields the best results.

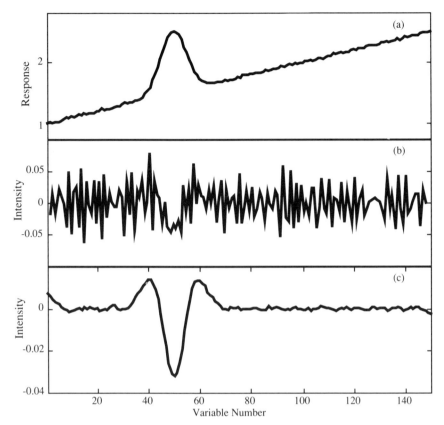

Figure 3.16. Comparison of calculating the second derivative of a sample vector with a linear baseline (*a*), using a simple difference (*b*), and the Gorry method with a window width of 11 (*c*).

An example from near-infrared reflectance spectroscopy is used to illustrate the application of derivative preprocessing. This analytical technique was considered for sorting recycled plastic containers (see Sections 4.2.1.2 and 4.3.1.2). Preprocessing was applied to minimize the differences between the spectra of containers of the same material. Figure 3.20*a* shows the spectra of the containers made of polyethylene. When using reflectance spectroscopy, wavelength-dependent scattering results in baseline variations which negatively impacts the analysis. The second derivative spectra shown in Figure 3.20*b* are more similar to each other than are the original spectra. The remaining differences are from variations in pathlength (see the normalization example, Section 3.1.1) as well as chemical variations between the samples.

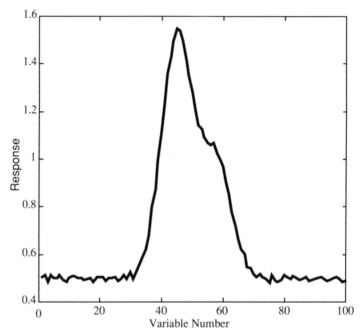

Figure 3.17. Sample vector to be used to demonstrate the effect of different window widths on the calculation of the derivatives. See Figures 3.18 and 3.19 for the application of the first and second derivative preprocessing, respectively.

3.1.4.3 MULTIPLICATIVE SCATTER CORRECTION

Multiplicative scatter correction (MSC) is a preprocessing tool developed to correct for the significant light-scattering problems in reflectance spectroscopy (Geladi et al., 1985; Martens and Næs, 1989). It has been found to be of more general use and is therefore sometimes referred to as multiplicative signal correction. When using MSC, one assumes that the variable number dependence of scattering or baseline signal is different from that of the chemical information. One advantage of this approach over the derivative methods is that the preprocessed spectra resemble the original spectra, which can aid in the interpretation. See Martens and Næs, 1989 for a mathematical description of the method.

An example of MSC preprocessing is shown in Figure 3.21(*a*) where the top graph displays a series of near-infrared reflectance spectra of solid polymer materials. The differences between spectra in the 1100- to 1300-nm region is smaller than the 1700- to 2500-nm region and this linear baseline can mask the signal of interest. After MSC preprocessing (Figure 3.21*b*), there is more variation in the peak centered at 2450 nm relative to the other regions. This is encouraging, but the advantage of this must be judged by performing the primary analysis on these preprocessed data.

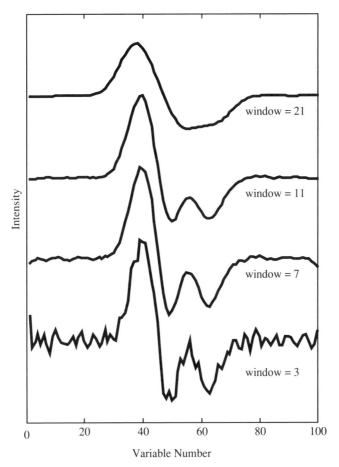

Figure 3.18. Comparison of using the Gorry method with varying window widths to calculate the first derivative of the data shown in Figure 3.17. (The offset was added for clarity.)

This concludes the discussion of baseline correction methods; a summary and recommendations are provided in Table 3.4.

3.2 PREPROCESSING THE VARIABLES

The second set of preprocessing tools discussed are those that operate on each variable. Table 3.5 lists the two methods discussed: mean centering and variable weighting. Mean centering is a common preprocessing tool that is applied to account for an intercept in the data. Variable weighting is used to emphasize some variables over others to increase their influence on the model.

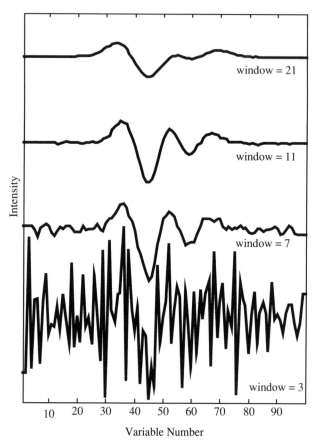

Figure 3.19. Comparison of using the Gorry method with varying window widths to calculate the second derivative of the data shown in Figure 3.17. (The offset was added for clarity.)

3.2.1 Mean Centering

Mean centering a variable is accomplished by subtracting the mean of that variable vector from all of its elements. (The elements in the variable vector correspond to the intensities for the same variable over different samples.) Performing the mean centering over multiple variables results in the removal of the mean sample vector from all sample vectors in the data set.

To illustrate mean centering, near-infrared spectra of organic mixtures are plotted in Figure 3.22a, with the mean-centered spectra shown in Figure 3.22b. The mean spectrum (overall shape) has been removed from all the spectra and the relative differences in intensity at the various wavelengths are easier to discern.

Data sets are often mean centered to account for an intercept term in regression models. To decide if mean centering is necessary, compare the results when the analysis is performed using centered and noncentered data.

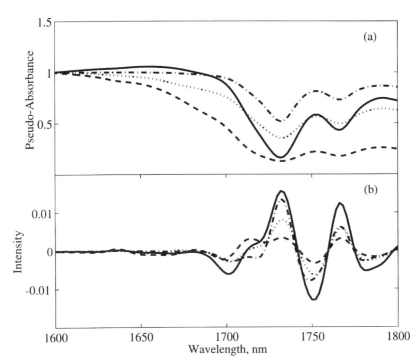

Figure 3.20. Near-infrared reflectance spectra of recycle polyethylene containers before (*a*) and after (*b*) calculating the second derivative using the method of Gorry.

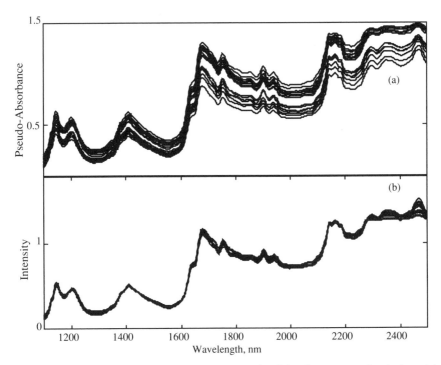

Figure 3.21. Near-infrared reflectance spectra of solid polymer samples before (*a*) and after (*b*) preprocessing using multiplicative scatter correction.

TABLE 3.4. Summary of Baseline Correction Tools

Method	Summary and Recommendations
Explicit modeling	Use if a functional form and baseline points can be identified.
Derivatives	A good general tool that does not rely on identification of baseline points. The methods of Savitzky–Golay or Gorry are preferred over the simple difference or running mean methods.
MSC	Results in interpretable spectra (unlike derivatives). Performance is often comparable to derivatives.

Mean centering generally does not hurt and often helps; therefore, many analysts use centering as a default. The one situation where we always recommend mean centering is when performing principal components analysis (see Section 4.2.2).

3.2.2 Variable Weighting

It is always possible to weight each variable by multiplying all elements in that variable vector by any number (a weight). If the weights are chosen wisely, this preprocessing step can improve the analysis results. In this section, four weighting schemes are discussed. The goal of the first two methods, a priori information weighting and variable selection, is to emphasize some variables over others. In contrast, the objective of variance scaling and autoscaling is to place all variables on equal footing. The methods are discussed below, and a summary is given in Table 3.6.

3.2.2.1 A PRIORI INFORMATION WEIGHTING As with sample weighting, this approach should be driven by accurate a priori information. This information may come from theory, prior experience, or experimentation. For example, measured signal-to-noise data can be used to downweight spectral regions with relatively poor signal-to-noise.

3.2.2.2 VARIABLE SELECTION An extreme weighting scheme is to eliminate some variables from consideration by using a weight of zero. Selection of measurement variables can be driven by statistical methodology, discussed in Section 5.3.1, or known chemical information. For example, absorbance bands may be selected based on knowledge of the chemistry under investigation, or eliminated based on signal-to-noise considerations.

TABLE 3.5. Variable Preprocessing Tools

Method	Use
Mean centering	Accounts for an intercept in a calibration model.
Variable weighting	Emphasizes some variables over others, increasing their influence on the primary analysis.

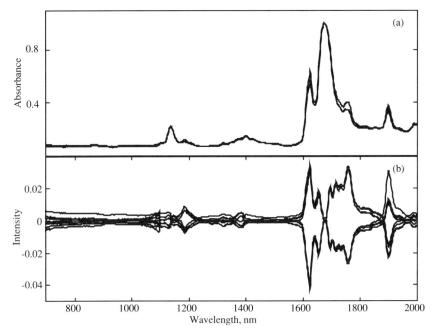

Figure 3.22. Near-infrared spectra of organic mixtures before (a) and after (b) mean centering.

Chemistry-based variable selection is applied while investigating the use of near-infrared reflectance spectroscopy for sorting recycle plastic containers (see Section 3.1.1). In this study the goal of variable selection was twofold: (1) minimizing the differences between spectra of containers of the same material, and (2) maximizing the differences between spectra of different types. Shown in Figure 3.23a are second-derivative spectra of the polyethylene samples in the 1100 to 2500 nm region. These samples are expected to have similar spectral features because they are made of the same material. Therefore, the regions with considerably different spectral features are eliminated (e.g., 2250–2500 nm). Other supporting data indicate that some regions are not effective for differentiating between plastic types. The final determination was that selecting the 1600- to 1800-nm region (Figure 3.23b) results in an improvement in the sorting of the plastic container types. This wavelength region includes the first overtones of the CH_3, CH_2, CH, and ϕ-CH transitions, and should therefore be useful for distinguishing the plastic containers.

3.2.2.3 VARIANCE SCALING Variance scaling is achieved by dividing each element in a variable vector by the standard deviation of that variable. The primary reason for variance scaling is to remove weighting that is artificially imposed by the scale of the variables. This is useful because many data

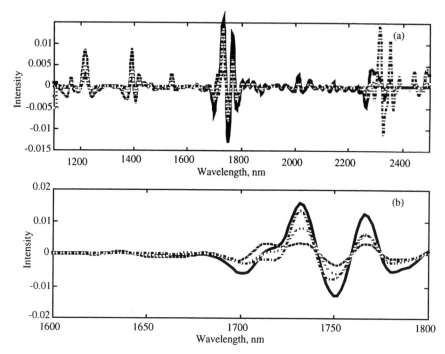

Figure 3.23. Near-infrared reflectance spectra with second-derivative preprocessing before (a) and after (b) region selection.

analysis tools place more influence on variables with larger ranges. This influence is arbitrary if the units of the variables are different. If one variable has units of grams it can be converted to kilograms, thus reducing its range by a factor of 1000. In practice, variance scaling is often performed in conjunction with mean centering, as discussed in the following section.

3.2.2.4 AUTOSCALING Autoscaling is the application of both variance scaling and mean centering (in either order). This preprocessing tool is used in the example discussed in Sections 4.2.2.2 and 4.3.2.2 where an array of seven Taguchi-type semiconductor sensors are considered for differentiating between organic vapors. Two measurements were made with each sensor: (1) the traditional, steady state intensity and (2) the maximum change in intensity over time. The resulting variables have intensity and slope units, respectively. The plot of the raw data shown in Figure 3.24a reveals that the slope measurements have much smaller values than the intensity measurements. The variability of the slope variables is masked, which can be seen by examining the slope variables on a different scale (Figure 3.24b). Autoscaling can be applied

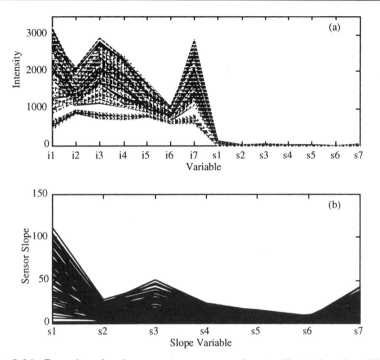

Figure 3.24. Raw data for the sensor-array experiments illustrating the different scales for (a) the intensity (i1–i7) and slope (s1–s7) measurements, (b) only the slope measurements.

to place the variables on a more equal footing (see Figure 3.25). The slope and intensity values are now unitless, so scale does not dominate the analysis.

Interpreting plots of autoscaled data is difficult because the units have been removed. Graphically it appears that autoscaling has removed much of the information because the patterns we are comfortable seeing are gone. However, this preprocessing step can in fact improve the analysis results.

This concludes the discussion of variable weighting methods; a summary and recommendations are provided in Table 3.6.

3.3 SUMMARY OF PREPROCESSING

Preprocessing, the second of the "Six Habits of an Effective Chemometrician," was discussed in this chapter. It is defined as any mathematical manipulation of the data before the primary analysis is performed, and is used to remove or reduce unwanted sources of variation for which the primary modeling method may not account. The tools were separated into two categories depending

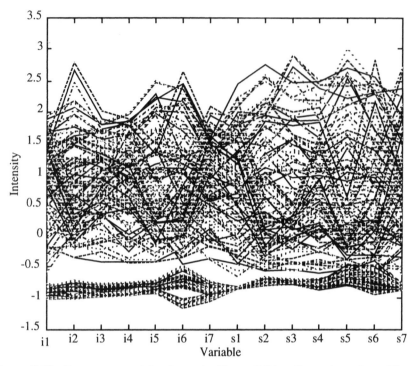

Figure 3.25. Sensor array data shown in Figure 3.24*a* after autoscaling. All variables are now unitless.

upon whether they operate on the samples or variables. Selecting the optimal tool often requires iteration between the primary analysis and the preprocessing step. While this iterative approach is common, it is best if the preprocessing is chosen because of a known feature about the measurement system and/or samples.

TABLE 3.6. Summary of Variable Weighting Methods

Method	Summary and Recommendations
A priori information	This scaling can be beneficial if reliable prior information concerning uncertainties or variable importance is available.
Variable selection	Knowledge of the analytical instrument and the chemistry of the system can be used to select variables in order to optimize model performance.
Variance scaling	Used to remove differences in units between variables.
Autoscaling	Accounts for an intercept in a calibration model and removes differences in units between variables. Useful when variables have different units; often not beneficial in other situations (e.g., spectroscopy).

REFERENCES

P. Barak, *Anal. Chem.*, **67**, 2758-2762 (1995).

N. Draper and H. Smith, *Applied Regression Analysis,* 2nd ed., Wiley, New York, 1981.

P. Geladi, D. McDougall, and H. Martens, *Appl. Spectrosc.*, **39**, 491-500 (1985).

P. A. Gorry, *Anal. Chem.*, **62**, 570-573 (1990).

P. A. Gorry, *Anal. Chem.*, **63**, 534-536 (1991).

P. R. Griffiths and J. A. de Haseth, *Fourier Transform Infrared Spectroscopy,* Wiley, New York, 1986.

D. Hoaglin, F. Mosteller, and J. Tukey, *Understanding Robust and Exploratory Analysis,* Wiley, New York, 1983.

I. Howe, D. H. Williams, and R. D. Bowen, *Mass Spectrometry: Principles and Applications,* 2nd ed., McGraw-Hill, UK, 1981.

K. Y. Hui and M. Gratzl, *Anal. Chem.*, **68**, 1054-1057 (1996).

W. H. Lawton and E. A. Sylvestre, *Technometrics,* **13**, 617 (1971).

H. Martens and T. Næs, *Multivariate Calibration,* Wiley, New York, 1989.

A. Savitsky and M. J. E. Golay, *Anal. Chem.*, **36**, 1627-1639 (1964).

4

Pattern Recognition

One of these things is not like the other, one of these things just doesn't belong . . .
—*Sesame Street*

Humans are very good at perceiving similarities and differences between objects of different shapes. For example, discriminating a square from a circle is learned by very young children through games and toys. For many young Americans, the first formal training in pattern recognition is at the hands of Big Bird and company on Sesame Street—finding which shape doesn't belong. The goal of pattern recognition in analytical chemistry is very similar: finding similarities and differences between chemical samples based on measurements made on the samples. For example, assume the pH, temperature, and density have been measured on four samples and the goal is to determine which samples are similar. Examining the resulting hypothetical measurement data in Table 4.1, the conclusion is that samples 1 and 4 are similar, as are samples 2 and 3.

Human pattern-recognition skills allow for the differentiation between rows with simple data matrices. However, this ability is limited to matrices with small numbers of rows and/or columns. For example, the data in Table 4.2 have a fairly simple mathematical relationship, but the human eye cannot see this. (These data are examined in more detail in Section 4.2.2.1.)

Because of this well-developed ability to perceive shapes, chemists often use pictures to present their data. For example, in spectroscopy, a spectrum is plotted as a continuous curve rather than represented in tabular form. The human eye can perceive the presence or absence of peaks, and interpretations are made accordingly. Using the computer to "perceive" these shapes or to enhance recognition abilities is the goal of pattern recognition.

To understand how the computer performs this task, one must first understand how the computer "views" chemical data. Although humans are better suited to interpret plots of data, the computer excels at the manipulation of tables of numbers. This distinction is shown in Figure 4.1, where three spectra are displayed using the chemist (top) and the computer (bottom) views. Each row in the matrix represents a different sample and each column is associated with a different wavelength.

TABLE 4.1. Hypothetical Measurement Data for Four Samples

Sample	pH	Temperature (°C)	Density (g/mL)
1	5	20	1.1
2	7	80	0.8
3	7	80	0.8
4	5	20	1.1

TABLE 4.2. Hypothetical Measurement Data for Many Samples

Sample Number	Measurement Variable 1	Measurement Variable 2	Measurement Variable 3
1	1.2036	0.3514	0.5336
2	1.5747	2.7024	1.3651
3	1.5610	0.8800	1.3329
4	2.4709	2.1856	1.3781
5	2.1031	3.3598	1.1370
6	3.1262	2.1694	2.4536
7	3.5547	3.5375	1.8349
8	2.8297	3.9520	2.0272
9	3.4619	2.0095	2.3668
10	4.1251	3.7992	3.2149
11	4.3519	4.7840	3.1682
12	4.3726	5.6372	3.1923
13	4.1362	5.5989	3.8251
14	5.1222	3.2704	3.2974
15	4.7702	4.3310	3.3547
16	4.9319	5.0789	3.7546
17	5.0776	5.6956	4.2358
18	4.5237	7.5085	3.8541
19	6.2747	3.1689	3.5633
20	6.2657	3.6446	3.9434
21	6.2770	4.7385	4.3812
22	5.6613	6.4239	4.7231
23	6.0396	6.6905	4.6699
24	7.1459	5.0001	4.4495
25	7.2072	6.5147	4.7595
26	6.5666	7.1129	5.3562
27	6.9216	8.5184	5.7890
28	8.3753	6.0726	5.5092
29	7.7007	7.1723	5.3151
30	8.5584	7.9795	5.6718
31	7.3148	9.0592	6.3253
32	8.7246	8.6834	6.5414
33	9.0110	8.7975	6.4387
34	9.7705	9.2219	7.0289
35	9.5646	9.9156	8.1231

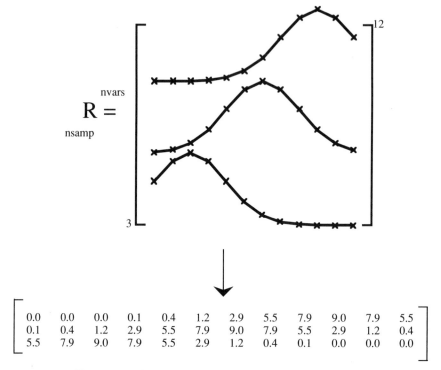

$$R = \begin{array}{c} \text{nvars} \\ \\ \text{nsamp} \end{array}$$

$$\begin{bmatrix} 0.0 & 0.0 & 0.0 & 0.1 & 0.4 & 1.2 & 2.9 & 5.5 & 7.9 & 9.0 & 7.9 & 5.5 \\ 0.1 & 0.4 & 1.2 & 2.9 & 5.5 & 7.9 & 9.0 & 7.9 & 5.5 & 2.9 & 1.2 & 0.4 \\ 5.5 & 7.9 & 9.0 & 7.9 & 5.5 & 2.9 & 1.2 & 0.4 & 0.1 & 0.0 & 0.0 & 0.0 \end{bmatrix}$$

Figure 4.1. Graphical and matrix representation of data.

Human versus computer pattern recognition can be understood using the example in Figure 4.2, where a matrix with two rows and two columns is shown in the upper left corner. If the rows of the matrix represent two spectra, a typical graphical representation is to place the measurement axis along the x axis and the intensities along the y axis. This results in a spectral representation of the type found at the top of Figure 4.1. In Figure 4.2 a quite different graphical representation is shown. This plot represents what is termed the row space of the matrix because the rows *reside* in this space. Each row is plotted as a point in a Cartesian coordinate system where the axes are defined by the columns. For example, the first row of the matrix **R** is plotted at (2,3) and the second row is plotted at (1,1). Now assume a third measurement is made on each sample. This corresponds to adding a third column to the matrix **R** in Figure 4.2. Plotting this requires adding a third axis to the plot, as shown in Figure 4.3.

Problems found in chemistry commonly are comprised of data sets with many measurements made on many samples. For example, a typical spectroscopic project might yield a data set with 30 rows (samples) and 700 columns (wavelengths). Using the definition described above, the row space of this matrix contains 30 points plotted in a 700-dimensional space. While it is not possible to construct this plot, the concept is simply an extension of

$$R = \begin{bmatrix} 2 & 3 \\ 1 & 1 \end{bmatrix}$$

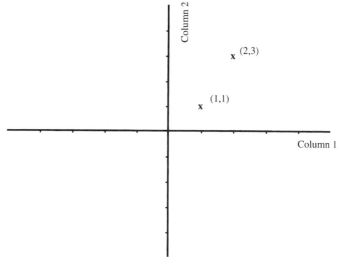

Figure 4.2. The row space plot of a matrix with two columns.

$$R = \begin{bmatrix} 2 & 3 & 4 \\ 1 & 1 & 3 \end{bmatrix}$$

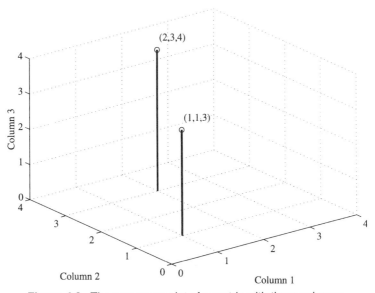

Figure 4.3. The row space plot of a matrix with three columns.

the examples discussed above. As the number of samples increases, so does the number of rows in the matrix. In row space, the number of points on the graph increases, because each point represents one sample. Likewise, the number of axes in the coordinate system increases as the number of columns (measurement variables) increases.

The next concept that must be understood is that of clustering of points (samples) in row space. Consider a simple pattern-recognition example in which the pressure (P) and temperature (T) of a batch reactor is measured during 10 runs. The data are tabulated in a matrix with the samples (batches) in rows and measurements (T,P) in the columns. The resulting matrix has 10 rows and 2 columns. Assume the collection of 10 points in the two-dimensional row space form two distinct clusters of points as seen in Figure 4.4. By definition, the points that are close to each other in the row space have correspondingly similar numbers in the matrix. This correspondence between numerical similarity and closeness in row space holds for all matrices, regardless of the number of columns or rows. It is this relationship that allows the computer to define numerical measures of similarity between rows (samples) in a matrix. Furthermore, when appropriate care is taken in selecting the measurement system, numerical similarity will reflect chemical similarity.

Two other points must be made relative to how the distances are measured. The first is that "distance" needs to be defined. A common metric used in Cartesian coordinate systems is Euclidean distance, where the distance between two points (x_1, y_1, z_1) and (x_2, y_2, z_2) in three-dimensional space is calculated as

$$\text{Distance} = \sqrt{(x_1 - x_2)^2 + (y_1 - y_2)^2 + (z_1 - z_2)^2} \qquad (4.1)$$

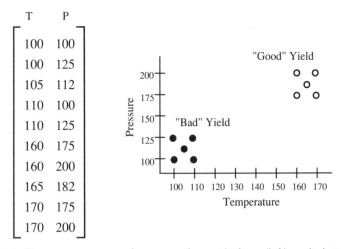

Figure 4.4. The temperature and pressure in matrix form (left) and plotted in row space (right).

However, the Euclidean distance is only one of many distance measures that can be used. For example, it may be desirable to weight certain measurement variables (dimensions) that are more reliable. This can be achieved by applying different weights to the terms in parentheses in Equation 4.1.

The second point to note is that none of these distance measures necessarily have any physical meaning. They are simply measures of proximity that are interpreted as similarity. Therefore, in all of the pattern-recognition techniques it is the relative distance between points (i.e., is point A closer to B than it is to C) that is important.

Consider again the example shown in Figure 4.4, but assume there is information as to whether the yield for each batch is "good" or "bad". The good and bad yield batches are clustered together in row space with the poor yields at low temperature and pressure. When the dimensionality of the problem increases, graphs cannot be made and, therefore, the human eye cannot be used to discern the patterns. Assume for the example given above that in addition to the temperature and the pressure, the stirring rate, catalyst concentration, and catalyst form were recorded. Directly viewing similarities or differences of operating conditions of the "good" and "bad" yield samples is not possible because the row space is five dimensional. This is where the tools described in this chapter become useful.

This example also introduces a concept that is used throughout this chapter—*class* membership. A class is defined as a collection of samples that are *defined* as being similar. This does not necessarily mean the samples will be close together in row space. For example, two classes can be defined as women (class 1) and men (class 2). The hair and eye color of the entire population (males and females) are measured. Distinct groupings of males and females would not be seen in the row-space plot because appropriate measurements have not been made to distinguish between these two classes. As analytical chemists, the *intent* is to choose measurements so that classes are discernible using pattern-recognition techniques. This illustrates the importance of chemical knowledge when choosing an analytical measurement technique. To be able to distinguish between samples, the chemist must choose a measurement technique that results in different values for chemically different classes of samples. This allows for differentiation because the classes will be far apart in row space.

Supervised versus Unsupervised Pattern Recognition

In some situations the class membership of the samples is unknown. For example, an analyst may simply want to examine a data set to see what can be learned. Are there any groupings of samples? Are there any outliers (i.e., a small number of samples that are not grouped with the majority)? Even if class information is known, the analyst may want to identify and display natural groupings in the data without imposing class membership on the samples. For example, assume a series of spectra have been collected and the goal is to identify any outliers. Pattern-recognition tools can accomplish this even if unique features of the outlying spectra are difficult to see using conventional plots. Class membership is not imposed on any of the samples, and the techniques applied are termed unsupervised pattern recognition (see Section 4.2). The goal is to increase the understanding of the data set by examining the natural clustering of the samples.

In other applications of unsupervised methods, the class information is known or suspected, but is not initially used. For example, suppose a sensor array has been used to obtain measurements on a collection of organic solvents. Unsupervised methods can be used to find natural groupings of solvents. By examining the groupings, chemical descriptors can be postulated that explain why different solvents are clustered together. Some potential descriptors include functional groups, molecular weight, and degree of saturation. This information can be used to determine what types of solvents the sensor array can classify. Additionally, understanding the chemical driving forces behind the clusters can give insight into the physical interactions between the solvents and sensors. This, in turn, can lead to the further development of the sensor array.

In contrast to unsupervised methods, supervised pattern-recognition methods (Section 4.3) use class membership information in the calculations. The goal of these methods is to construct models using analytical measurements to predict class membership of future samples. Class location and sometimes shape are used in the calibration step to construct the models. In prediction, these models are applied to the analytical measurements of unknown samples to predict class membership.

An example of a supervised model is one developed for classifying material in a holding tank as raw material A, B, or C. Measurements are made on samples from each class (materials A,B,C) and a model is derived that distinguishes between the classes. The resulting model is used to verify the identity and integrity of incoming raw materials before use. This is possible if the analytical measurements are sufficiently sensitive and selective to detect differences between the raw materials.

4.1 DECISION TREE

The decision tree in Figure 4.5 can be used to choose the appropriate pattern recognition method(s) for a given problem. The first question to ask is whether the goal is to develop a model that will be used to predict the class membership of future samples, or is it to simply examine the similarities and differences between the samples? If the latter, the unsupervised pattern-recognition techniques described in Section 4.2 are appropriate. These methods are termed unsupervised because the algorithms are not guided by a priori class membership information. The two unsupervised techniques discussed in this chapter are Hierarchical Cluster Analysis (HCA) and Principal Components Analysis (PCA).

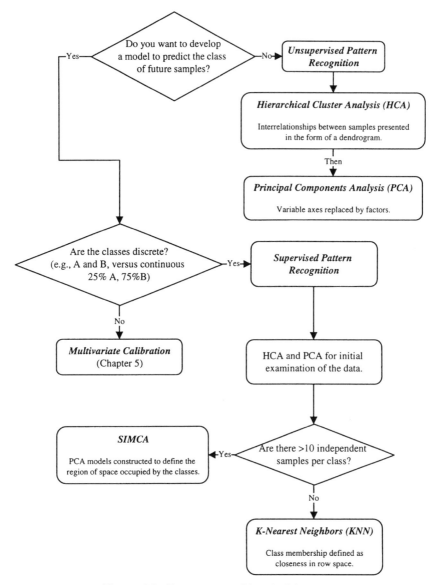

Figure 4.5. Pattern recognition decision tree.

If the goal is to develop a predictive model, a training set with known class memberships is required to construct the model and the techniques are termed "supervised." The techniques are designed to operate on data sets with discrete class membership. For example, oranges belong to the "fruit" class whereas broccoli belongs to the "vegetable" class. If the measurements on the samples in a data set are continuous in nature (e.g., concentrations of component A), it is usually more effective to use one of the multivariate calibration tools found in Chapter 5 to solve the problem. Having decided that a supervised pattern-recognition method is the correct tool, our recommendation is to begin by examining the data using the unsupervised tools. By applying unsupervised methods first, the data are examined with the minimum amount of potential bias. Additionally, the unsupervised methods are good for examining the feasibility of supervised methods and identifying outliers that can adversely affect the model construction. Having removed obvious outliers and gained confidence in the feasibility of the classification, a predictive model can be derived. The two supervised techniques discussed in this chapter are K-Nearest Neighbor (KNN) and Soft Independent Modeling of Class Analogy (SIMCA). The KNN technique is typically used if there are not many samples per class. This is because no assumptions about the size or shape of the class are necessary to construct a model. In SIMCA, a PCA model is constructed for each class, and so it is necessary that there are sufficient samples to construct an adequate model.

4.2 UNSUPERVISED LEARNING

The goal of unsupervised learning is to evaluate whether clustering exists in a data set without using class membership information in the calculations. The row space is examined to reveal any natural groupings of the samples as defined by the measurements. Examples of what can result from an unsupervised pattern-recognition analysis are shown pictorially in Figure 4.6. There are only two possible outcomes: either multiple clusters are present in the data (Figure 4.6a) or the samples all fall within one cluster (Figure 4.6b).

The presence or absence of clusters is dependent on the measurements chosen and the samples in the data set. The fact that only one cluster appears in Figure 4.6b may not be what is expected. If more than one cluster was expected, further investigation might involve changing expectations (i.e., the samples really only belong in one class), choosing different measurements (i.e., the current measurements do not have adequate discriminating power), or improving the data [i.e., preprocessing might be needed (see Chapter 3)].

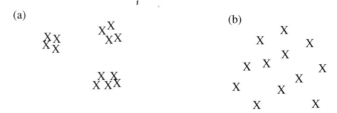

Figure 4.6. The two results of unsupervised learning are (a) multiple clusters are present and (b) the data are all within one cluster.

Figure 4.7. Additional information for the interpretation of the result shown in Figure 4.6a; (a) clustering corresponds to known information or (b) clustering is not related to known information.

When class information is available, it is of interest to compare this with the natural groupings. For the example in Figure 4.6a, assume class information is available. When applying this information, the clustering might or might not relate to the class information (Figures 4.7a and b, respectively). The latter indicates a disconnect between the measurements that were chosen and the expected groupings. Further investigation is warranted, which might include postulating different classes by examining the clusters to determine if there is a reasonable chemical explanation for the groupings, obtaining additional class information through further experimentation, or using principal components analysis to understand which variables contributed to the clustering (see Section 4.2.2).

As depicted in these examples, unsupervised methods are useful tools for presenting the natural clustering of samples within the data set. This view of the data, coupled with an understanding of the measurement system, often aids in elucidating the physical reasons for the presence or absence of clustering in the data.

4.2.1 Hierarchical Cluster Analysis (HCA)

Hierarchical cluster analysis (HCA) is an unsupervised technique that examines the interpoint distances between all of the samples and represents that information in the form of a two-dimensional plot called a dendrogram. These dendrograms present the data from high-dimensional row spaces in a form that facilitates the use of human pattern-recognition abilities.

To generate the dendrogram, HCA methods form clusters of samples based on their nearness in row space. A common approach is to initially treat every sample as a cluster and join closest clusters together. This process is repeated until only one cluster remains. Variations of HCA use different approaches to measure distances between clusters (e.g., single vs. centroid linking, Euclidean vs. Mahalanobis distance). The two methods discussed below use single and centroid linking with Euclidean distances.

The single-link approach links clusters based on the distance between the "nearest neighbors" contained in the clusters, as illustrated in Figure 4.8. In step i the samples are initially in five individual clusters, and samples A and B

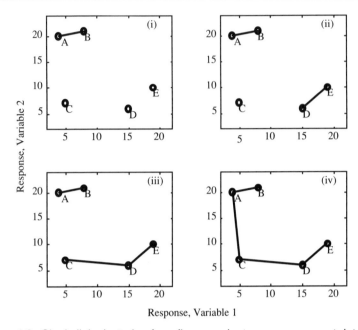

Figure 4.8. Single link clustering for a five-sample, two-measurement data set.

are the closest clusters. Because these are the closest two clusters, they are joined as indicated by the solid line. Once samples A and B are joined into one cluster, the procedure is repeated with the remaining four clusters. In step ii, clusters D and E are the closest to each other, and are linked. Then in step iii, cluster C is linked with the DE cluster, because the distance from C to D is smaller than the distance between any other pair of clusters. Finally, in step iv, the two remaining clusters are joined, resulting in a single cluster.

The clustering information in these plots can be represented in a two-dimensional dendrogram regardless of the dimensionality of the row space (which is the number of measurement variables). The dendrogram for the data in Figure 4.8 is shown in Figure 4.9. The samples are listed on the left-hand side of the graph, and the vertical lines indicate which samples are in a cluster. The x axis is a measure of distance between clusters, and the locations of the vertical lines correspond to the distances between two joined clusters. For example, the vertical line joining clusters A and B indicates the distance between these points is approximately four units. (Many software packages convert the x axis to units of "similarity" ranging from 0 to 1, where identical clusters have a similarity equal to 1.)

Moving from left to right on the dendrogram, the number of classes decreases to one, and the distance between the clusters increases. Comparing Figures 4.8 and 4.9 illustrates how the dendrogram is constructed. Clusters A and B are the closest in Figure 4.8, and the vertical line connecting these two clusters in Figure 4.9 is the furthest to the left. The next closest clusters are D

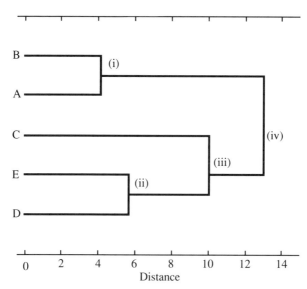

Figure 4.9. Dendrogram of the data in Figure 4.8 using the single-link clustering method.

and E, and the vertical line connecting these points in the dendrogram is the next vertical line to the right. Although these two figures contain largely the same information, it is important to note that some geometric information is lost when going from Figure 4.8 to Figure 4.9. For example, when C is joined to DE, the dendrogram does not indicate whether it was closer to D or E.

The second approach used for linking clusters is termed the centroid link method. This method joins the centroids of the clusters instead of the nearest neighbors. The centroid is calculated as the average of the points within a cluster. Clustering using this approach with the same data as in Figure 4.8 is shown in Figure 4.10.

The first two steps are identical to the single-link method because they involve joining clusters containing single points. (The centroid of a single point cluster is itself.) However, subsequent steps involve linking the centroid of existing multipoint clusters. In step iii, C is joined to the centroid of the DE cluster. The last cluster is formed in step iv by joining the centroid of the AB cluster to the centroid of the CDE cluster.

Figure 4.11 is the dendrogram corresponding to the data shown in Figure 4.10. Comparing this dendrogram to the single-link dendrogram (Figure 4.9) reveals very small differences. (Note that the samples are in a different order along the y axis in Figures 4.9 and 4.11; this is simply a function of the plotting algorithm.) The distances associated with the first two clusters formed (AB and DE) are identical in the two dendrograms. The order in which the remaining clusters are formed is also identical. The only meaningful differences between the dendrograms is that the distances are slightly different for steps

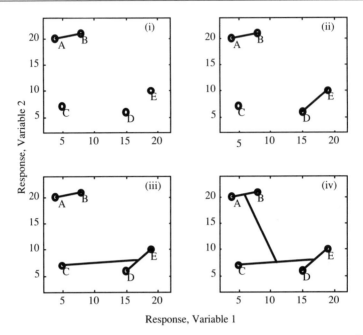

Figure 4.10. Centroid link clustering for a five sample, two measurement data set. The four steps to construct the dendrogram are illustrated by i–iv.

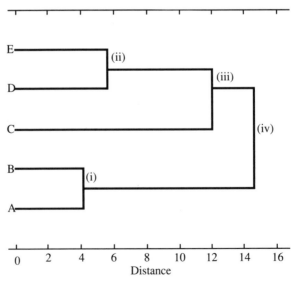

Figure 4.11. Dendrogram of the data in Figure 4.10 using the centroid-link clustering method.

iii and iv. The distance between clusters C and DE is approximately 10 units using the single-link method and is almost 12 using the centroid method. This is also apparent when examining step iii in the row space plots, Figures 4.8 and 4.10.

The single and centroid linking methods do not always join clusters in the same order as in the previous example. Figure 4.12 shows an example where these methods yield different dendrograms. If the single-link method is used, the clustering proceeds as follows (solid lines in Figure 4.12):

Step i	Join clusters A and B (distance = 6.0)
Step ii	Join clusters C and D (distance = 7.4)
Step iii	Join clusters AB and CD (distance = 7.6)

This results in the dendrogram found in Figure 4.13a. If the centroid-link method is used on these same data, the clustering scheme is as follows (dashed lines in Figure 4.12):

Step i	Join clusters A and B (distance = 6.0)
Step ii	Join clusters C and AB (distance = 7.0)
Step iii	Join clusters D and ABC (distance = 10.4)

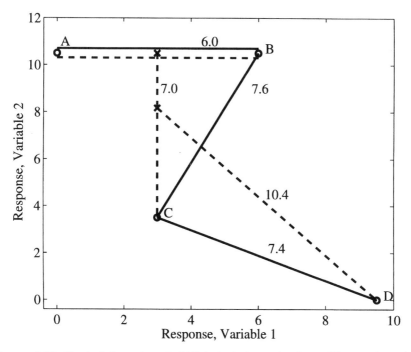

Figure 4.12. Single-link and centroid-link techniques lead to different clustering schemes. Solid, single link; dashed, centroid link.

The resulting dendrogram is found in Figure 4.13*b*.

It is not possible to make a general statement as to whether the single- or centroid-link approach will perform better. The centroid method results in a better measure of the distance between clusters when the clusters are well separated. This is because it measures the distance between the centers of the clusters rather than the distances from the edges as in single link. On the other hand, the centroid method is sensitive to outliers because these unusual points

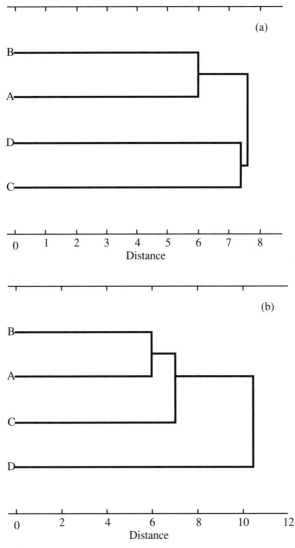

Figure 4.13. Dendrograms for data shown in Figure 4.12: (*a*) single link; (*b*) centroid link.

adversely influence the calculation of the centroid of a cluster. With well-separated classes, both methods give similar results. Because the goal is to learn as much as possible about the data, the recommendation is to use both methods and compare the results. For brevity, only the single link results are presented in the examples that follow.

4.2.1.1 HCA Example 1   A data set comprised of two measurements made on 42 samples is examined. The goal is to use HCA to learn how this collection of samples is clustered in row space. The discussion follows the "Six Habits of an Effective Chemometrican" which are detailed in Chapter 1.

Habit 1. Examine the Data

For this example, because there are only two variables, it is possible to plot the row space (variable 2 vs. variable 1) and look for patterns. Generally this plot is not available because most data sets are higher dimensional (>3). Therefore, it is not plotted here.

The graphical representation that is universally available is shown in Figure 4.14. This plot is examined for features such as sample outliers, the need for preprocessing, questionable variables, and other patterns. In this case, there appears to be three clusters and two additional "unusual" samples. These two samples are within the scale of the entire data set, but have a different response pattern.

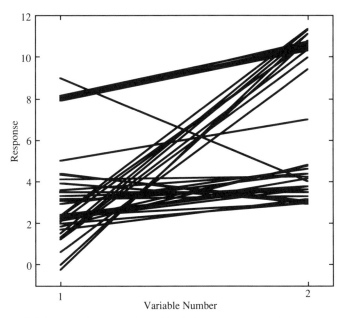

Figure 4.14. HCA example 1 raw data. Each line represents the measurements for a different sample.

Habit 2. Preprocess as Needed

Consider autoscaling if the measurements are known to have different units (see Chapter 3). It is difficult with this example to determine the need for preprocessing by examining the plot. In more typical applications, there are more than two measurement variables and the need for preprocessing is easier to discern. Keep in mind that preprocessing will usually change the relative position of the samples in row space. Therefore, exercise caution so as not to remove pertinent information by preprocessing. With this data set, no preprocessing was performed.

Habit 3. Estimate the Model

The information submitted to the computer program includes the preprocessed data, the metric for classification (e.g., Euclidean distance), and the choice of clustering method (e.g., single link). Intercluster distances are calculated and used to construct a dendrogram.

Habit 4. Examine the Results/Validate the Model

Two diagnostic tools are discussed below and a summary is found at the end of the section in Table 4.3. All of the HCA diagnostic tools are used to study the relationships between the samples. When interpreting the dendrogram, pose the following questions about the samples: (1) Does the clustering (if any) make sense? (2) What can be learned about the relationship between samples? (3) Are any further experiments suggested? (4) Are there any unusual samples or grouping of samples (outliers, something unexpected)?

Dendrogram Without Class Labels (Sample Diagnostic): Even if class information is available, it is advisable to examine the data first without the class labels to reduce bias in the interpretation. The dendrogram for this example is shown in Figure 4.15. The sample identities (indicated for now by X) are listed on the left-hand side of the graph, and the vertical lines indicate which samples constitute a cluster.

There are three large clusters of samples that are joined at a relatively large distance. The preliminary conclusion is therefore that three classes of samples are present in the data set. These clusters are identified in the dendrogram as "cluster 1," "cluster 2," and "cluster 3." Two other samples appear to be different from the others and are labeled "unusual sample 1" and "unusual sample 2." They are considered unusual because they are solitary samples that are joined with a cluster at relatively large distances

Referring back to the main clusters, it can be seen that the samples within cluster 2 are connected by vertical lines with small distance values relative to the other clusters. This is an indication that samples within this cluster are more similar to each other (i.e., the interpoint distances are smaller) than are the samples within the other clusters.

Dendrogram With Class Labels (Sample Diagnostic): Now assume it is known that there are five classes (A–E). The dendrogram with the class labels for this example is shown in Figure 4.16. With this class information, it be-

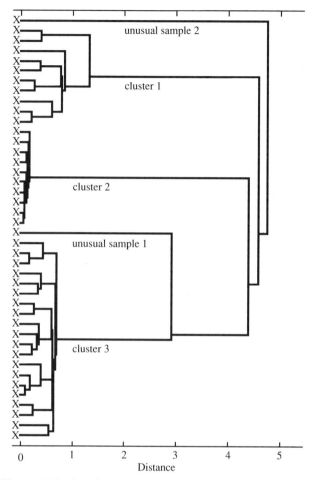

Figure 4.15. Dendrogram for the data plotted in Figure 4.14.

comes clear that what initially was considered cluster 3 actually consists of two overlapping classes (C and D). That is, some samples in class C are closer to some class D samples than other samples in their own class. Additional measurements would need to be considered to successfully separate all of these classes.

Referring again to Figure 4.16, unusual sample 1 is labeled as a member of class A. This should be investigated because all other samples from class A form one cluster (cluster 1). At least three explanations are possible for this outlying sample: (1) an error occurred when collecting the measurement data, (2) the sample was mislabeled, or (3) class A is diverse and only one sample captured another source of variation. If the sample is mislabeled, it should be contained within another cluster. However, it is linked to the nearest cluster

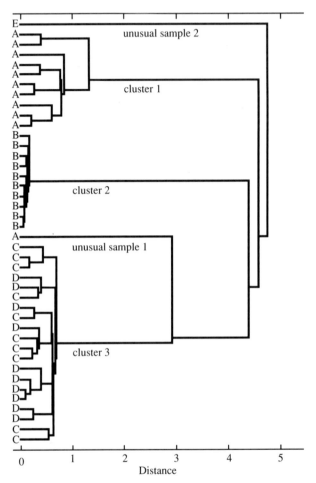

Figure 4.16. Dendrogram for the data plotted in Figure 4.14, with class identification labels.

(cluster 3) at a distance of ~3 while all other members of cluster 3 have interpoint distances of less than 1. Mislabeling, therefore, does not appear to explain this unusual sample. If this sample is really a member of class A, either there is another problem with this sample or additional work is needed to fully characterize class A.

To further investigate this sample, it is helpful to reexamine the raw data plot paying careful attention to class A samples. Given the separation of unusual sample 1 from the other class A samples in the dendrogram, spectral differences should be easily identifiable. In Figure 4.14, the class A samples are clustered with average response equal to (1,10), while unusual sample 1 has a response of (5,7). The challenge remains to determine why the unusual sample has distinguishing features. Begin by interpreting the spectral features, looking for entries regarding this sample in laboratory notebooks, studying the chemistry, and/or repeating the analysis on the sample.

Before using the class-membership information, unusual sample 2 might have been considered an outlier. However, it no longer appears to be unusual because it is the only member of class E in the data set. Be aware that it is not possible to characterize the dispersion of the class with only one sample.

Summary of Validation Diagnostic Tools for HCA: The row space plot for this example is shown in Figure 4.17 with the class information included (remember that in most cases it is not possible to obtain this plot). This plot verifies the conclusion drawn from the dendrogram, that is,

1. Samples from class B are very closely clustered, and are also well separated from samples from classes A, C, D, and E.
2. Samples from class A are well separated from the other classes, but are not as tightly clustered as B samples. This could be due to natural chemical variation or the quality of the laboratory preparation and measurements.

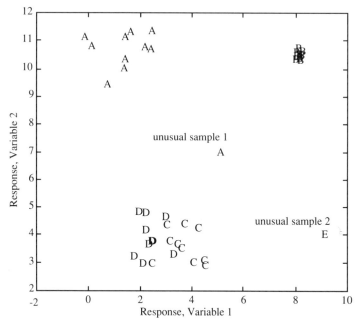

Figure 4.17. Row-space plot of data for Example 1 with the class information shown.

TABLE 4.3. Summary of Validation Diagnostics for HCA

Diagnostics	Description and Use
Model	No model diagnostics are discussed.
Sample	
Dendrogram without class labels	The dendrogram shows the closeness of samples in row space in the form of a two-dimensional graph.
Sample vs. distance (or similarity)	It is used to examine the similarities and differences between samples without imposing a priori information regarding class membership.
	Samples joined at small distances are similar based on the measurement system.
	Samples joined at large distances can indicate outliers.
Dendrogram with class labels [Class vs. distance (or similarity)]	Same description as Dendrogram without class labels above except the class membership is included.
	The class labels are used to verify that the clustering in the dendrogram corresponds with the known class information.
Variable	No variable diagnostics are discussed.

3. Samples from classes C and D are separated from classes A, B, and E, but not resolved from each other by the measurements used. Further experiments are needed to identify additional measurements to discriminate between these two classes.

4. The sample for class E is well separated from all other classes. More data must be collected to determine the dispersion of class E.

5. One of the samples that is labeled A is not similar to any other sample from its assigned class, nor is it similar to any other sample in the data set. It is unusual, but the dendrogram does not help assign a cause.

Table 4.3 summarizes the validation diagnostic tools discussed in this section. The first column in the table lists the name of each tool and the second column describes results from both well-behaved and problematic data.

Habit 5. Use the Model for Prediction

Habit 6. Validate the Prediction
The final two habits are not applied when using unsupervised methods.

4.2.1.2 HCA Example 2   The goal of the project described here is to use HCA to help determine whether near-infrared reflectance spectroscopy (NIR) can be used to facilitate the sorting of recycled plastic containers. (This data set is also discussed in Example 2 of KNN, Section 4.3.1.2.) Waste plastic containers were obtained and the reflectance spectra measured in the range 1100–2500 nm. The materials represented in the data set are listed in Table 4.4.

TABLE 4.4. Waste Plastic Samples used for NIR Feasibility Study

Plastic Type	Number of Samples
Polyethylene terephthalate (PET)	13
Polypropylene (PP)	3
Polystyrene (PS)	5
Polyvinyl Chloride (PVC)	4
Polyethylene (PE)	4

Habit 1. Examine the Data

The raw data (baseline corrected at 1600 nm) for all of the plastic samples plotted in Figure 4.18 reveals significant "within class" variation (the offset for the different plastic types was added for clarity). However, none of the samples have features that appear to be unusual given that the reflectance spectra are of unwashed plastic containers. The decision is to leave all samples in for further analysis.

Habit 2. Preprocess as Needed

The raw data plots and experience with near-infrared reflectance spectroscopy indicates the need for preprocessing to minimize the effects of scatter. However, to illustrate the effect of preprocessing on the HCA analysis, no preprocessing will be performed for the first pass through the data.

Habit 3. Estimate the Model

Single-link clustering and Euclidean distance are used in the calculations.

Habit 4. Examine the Results/Validate the Model

Dendrogram Without Class Labels (Sample Diagnostic): The dendrogram of the raw data shown in Figure 4.19 does not reveal five distinct clusters, as is desirable given that five plastics are represented. Although not shown here, the addition of the class membership information does not help resolve the overlap of the classes. Given no additional information about the problem, the conclusion would be that the current approach is not acceptable for sorting the plastic containers.

It is not surprising that distinct classes were not found because of the adverse effect of scatter on the spectra. The goal is to produce more similar looking spectra within each plastic type (i.e., reduce the within-class variation). Therefore, knowing which samples are made of the same material can also help in selecting appropriate preprocessing methods. Applying region selection, second derivative, and normalization produces the preprocessed data shown in Figure 4.20. (For details of the preprocessing of these data, see Section 4.3.1.2.) The within-class variation has been significantly decreased, while much of the between-class differences have been preserved (compare Figures 4.18 and 4.20). The dendrogram without class information for the preprocessed data is shown in Figure 4.21. Five clusters of samples are evident with no unusual samples. This is consistent with the number of known classes (plastic types).

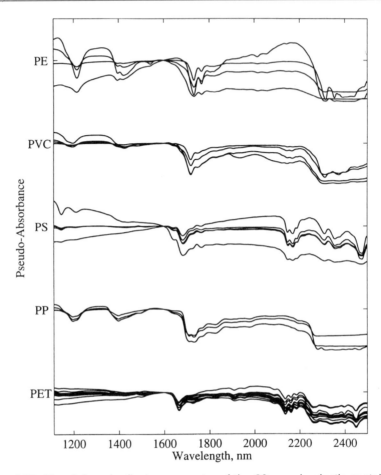

Figure 4.18. Near-infrared reflectance spectra of the 29 recycle plastic containers, baseline corrected at 1600 nm.

Dendrogram With Class Labels (Sample Diagnostic): Adding the class information (Figure 4.22) verifies that the five clusters correspond to the five plastic types. Other features that can be seen include:

1. All of the classes have approximately the same dispersion (size) and there is good separation between the classes. Four of the five clusters have almost equal dispersion, evidenced by the fact that the last sample to be joined to each group is at a distance of ~0.04.

2. The PET class has many samples that are very closely clustered (linked at a distance of ~0.1). This can also be seen in the plot of the preprocessed data (Figure 4.20).

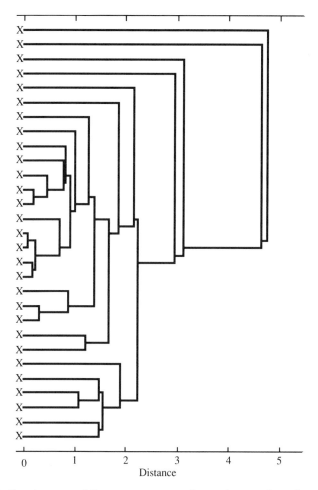

Figure 4.19. Dendrogram of the unpreprocessed recycle container data shown in Figure 4.18.

3. The two class that are most similar are PE and PS. They link together at the shortest distance (~0.16 units).

4. The PE class has the most within class variability; this can also be seen in the plot of preprocessed data (Figure 4.20).

As an exercise, the reader is encouraged to examine Figure 4.20 and identify the corresponding samples in the dendrogram in Figure 4.22. For example, each plastic type has one spectrum that is most different from the other samples. In all cases, they are the samples that are joined to their respective class at the largest distance.

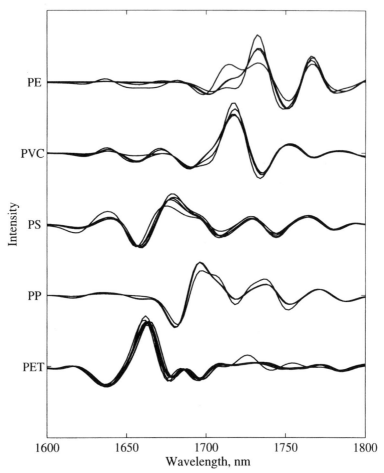

Figure 4.20. Near-infrared reflectance spectra of the 29 recycle plastic containers, preprocessed with a second derivative, variable selection, and normalization.

Summary of Validation Diagnostic Tools for HCA: The conclusion from this example is that NIR is capable of classifying containers made of the five materials investigated. The example also demonstrates how preprocessing can remove sources of variation that are not related to the differences between the plastics.

Habit 5. Use the Model for Prediction

Habit 6. Validate the Prediction
The final two habits are not applied when using unsupervised methods.

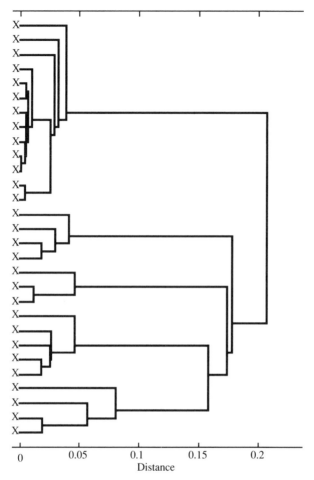

Figure 4.21. Dendrogram of the preprocessed recycle container data shown in Figure 4.20.

4.2.2 Principal Components Analysis (PCA)

As discussed in the introduction to this chapter, examining the row space of a matrix is an effective way of investigating the relationship between samples. However, this is only feasible when the number of measurement variables (columns) is less than three. Principal components analysis is a mathematical manipulation of a data matrix where the goal is to represent the variation present in many variables using a small number of "factors." A new row space is constructed in which to plot the samples by redefining the axes using factors rather than the original measurement variables. These new axes, referred to as factors or principal components (PCs), allow the analyst to probe matrices

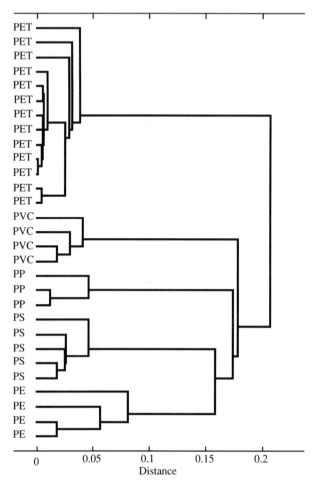

Figure 4.22. Dendrogram of the preprocessed recycle container data shown in Figure 4.20, with class-identification labels.

with many variables and view the true multivariate nature of the data in a relatively small number of dimensions. With this new view, human pattern recognition can be used to identify structures in the data.

Principal components analysis can be best understood using a simple two-variable example. With only two variables it is possible to plot the row space without the need to reduce the number of variables. Although this does not fully present the utility of PCA, it is a good demonstration of how it functions. A two-dimensional plot of the row space of an example data set is shown in Figure 4.23. The data matrix consists of two columns, representing the two measurements, and 40 rows, representing the samples. Each row of the matrix is represented as a point (O) on the graph.

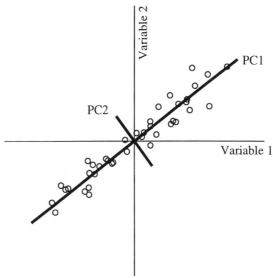

Figure 4.23. A row plot of data in a two-measurement system, with the first two principal component axes drawn in bold.

It is of interest to study the relationship between the samples in the row space; the distances between samples are used to define similarities and differences. In mathematical terms, the goal of PCA is to describe the interpoint distances (spread or variation) using as few axes or dimensions as possible. This is accomplished by constructing PC axes that align with the data.

Neither of the original variables in Figure 4.23 (variable 1 or 2) completely describes the variation in the data set. However, the first principal component (PC1 in Figure 4.23) is calculated such that it describes more of the variation in the data than either of the original axes. All principal components have the following properties:

1. The first principal component explains the maximum amount of variation possible in the data set in one direction. Stated another way, it is the direction that describes the maximum spread of data points. Furthermore, the percent of the total variation in the data set described by any principal component can be precisely calculated.

In this example, the first principal component describes 98% of the variation. Successive PCs can be estimated that describe a portion of the remaining variation; in this example, the second PC contains 2%. This illustrates another property of PCs, that is, successive PCs describe decreasing amounts of variation. Knowing the percent variation described is very important when interpreting the plots. For example, if close to 100% of the variation is described using two PCs, a two-dimensional plot can effectively be used to study the variation in the data set. However, a two-dimensional plot will not be adequate if only 20% of the variation is described—too much variation is missing. Therefore, conclusions based on examination of PC plots should be tempered by how much of the variation is captured.

2. Just as the sample has coordinates in the original row space (defined by the original variables), it also has coordinates with respect to the new PC axes. These can be found by drawing a perpendicular line from the sample to the principal component axes. In Figure 4.24, the dash–dot line shows the coordinates of one sample relative to the two original variable axes, and the dashed line shows the coordinates of that sample relative to the first principal component axes. The coordinates of the samples relative to the principal component axes are typically termed "scores."

3. Each PC is constructed from combinations of the original measurement variables. The extent to which a measurement variable contributes to a PC depends on the relative orientation in space of the PC and variable axes. For example, in Figure 4.23, PC1 is slightly more parallel to the variable 1 axis than to the variable 2 axis, which means that variable 1 contributes more to PC1 than does variable 2. For interpretation, it is useful to know which variables contribute most significantly to the individual principal components. These are the axes that are potentially the best at discriminating between samples.

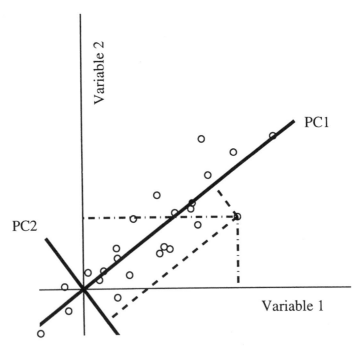

Figure 4.24. The first quadrant of the data in Figure 4.23 showing the coordinates of one point relative to the original axes (the dashed–dotted lines) and the principal component axes (the dashed lines).

In mathematical terms, the contribution of each axis to a principal component is the cosine of the angle between the variable axis and the principal component axis. If a principal component points exactly in the same direction as an individual variable, the angle between them is 0, and the cosine is 1. This indicates that the PC describes all of the variation in that variable axis. Similarly, if a principal component is perpendicular to an individual variable axis, the cosine is 0, indicating that none of the variation is contained in the PC. These cosine values are often termed "loadings," and can range from -1 to 1.

4. Excluding nonsignificant principal components can be used to filter noise from a data set. This is because the first principal components are calculated such that they describe the largest variation in the data set. Each PC also contains noise, but the noise is spread out amongst all of the PCs. The ratio of signal-to-noise is highest in the first PCs and decreases as subsequent PCs are calculated. The result is that nonsignificant principal components typically describe more noise than signal and, therefore, their exclusion filters noise from the data set.

5. The maximum number of PCs that can be calculated is the smaller of the number of samples or variables.

6. As used in chemometrics, principal components are orthogonal (perpendicular) to each other.

In the current example, the majority of the variation observed in two dimensions is described using one principal component (98%). The data essentially forms a line coincident with PC1 (with some scatter). This demonstrates the ability of PCA to more efficiently describe the variation in the data set.

For chemical applications, this is a very powerful tool. For example, assume 800 measurements are made per sample with a spectroscopic technique on a series of samples with three chemical components changing in concentration. While the spectra exist as points in an 800-dimensional space, the samples only occupy three dimension within that space (assuming Beer's Law is obeyed). This is because there are only three sources of variation in the data set—the varying concentrations of components. The spectra of a single analyte with varying concentrations lie on a line which passes through the origin (corresponding to a sample with zero concentration). The spectra of a second analyte with varying concentrations also lie on a line, but the direction of the line in row space is different. The spectrum of any sample containing these two components must lie on the plane formed by these two lines. Adding a third analyte defines a three-dimensional space in which all the samples lie. Principal components analysis is capable of presenting graphs of the three-dimensional space that contain all the variation, making it possible to examine the complete data set in one three-dimensional plot.

Consider the advantage of the principal component approach over the alternatives. One option is plotting combinations of three variables from the original 800 variables. This is not as comprehensive as with the PCA graphs because three individual variables typically will not capture the variation within the entire data set, because many of the variables often contribute significant variance to the data set (i.e., the cluster of samples in row space is often tilted with respect to the original measurement axes). Therefore, plotting single variables does not describe an adequate amount of variation. With PCA, each PC contains the variation from multiple original variables and therefore can more efficiently describe the variation in the data set.

Principal components analysis is most useful when the dimensionality of the measurement space is large (many columns), but where the samples reside in a small dimensional space (small *inherent* dimensionality). When three or fewer PCs describe a significant amount of the variation, one three-dimensional plot can be used to display the variation. In many chemical examples, the inherent dimensionality is relatively small compared to the number of measurement variables. It is limited by the number of physical phenomenon that were changing during data collection (this is discussed later). In PCA terms, the inherent dimensionality is defined as the number of principal components needed to describe a set of data to the noise level. For example, a line has an inherent dimensionality of 1 and a plane has an inherent dimensionality of 2.

The main challenge of PCA is determining the number of relevant principal components. This is equivalent to defining the amount of relevant variation that is contained in a data set. To complicate matters, how much variation described is enough is highly dependent on the problem at hand. If the noise level is 0.1%, using three PCs to describe 99.0% of the variation may not be adequate. However, three PCs describing 99.0% of the variation is too many if the known noise level is 10.0%. Choosing the correct number of PCs is important in order to view only relevant variation. Including too much noise by using too many PCs or excluding pertinent information by using too few PCs will adversely affect the interpretation of the data. Guidelines for determining the inherent dimensionality are presented in Section 4.2.2.1.

A number of constraints can limit the inherent dimensionality of a data set. For example, in Figure 4.25*a* there are only two measurements and therefore it is not possible for the data to occupy a space larger than two dimensions (i.e., the inherent dimensionality cannot be larger than the number of measurement variables). The number of samples can also limit the inherent dimensionality. In Figure 4.25*b* the inherent dimensionality can be no more than 1 because only two samples are measured (and two points define a line). More often in chemometrics, the chemistry is what limits the inherent dimensionality of a set of data. Figure 4.25*c* shows the result of measuring two variables on nine samples containing one chemical species. The concentration is varying and the measurements are linear with respect to concentration. The result is that the samples lie on a line which has an inherent dimensionality of 1 (associated with the one chemical species).

A variety of algorithms can be used to calculate the loadings and scores for PCA. A commonly employed approach is the singular value decomposition (SVD) algorithm (Golub and Van Loan, 1983, Chapter 2). A matrix of arbitrary size can be written as $\mathbf{R} = \mathbf{USV}^T$. The **U** matrix contains the coordinates of the

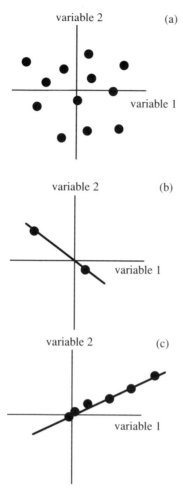

Figure 4.25. Samples plotted in row space where the dimensionality is limited by (a) the number of measurements, (b) the number of samples analyzed, and (c) the number of chemical components present in the system.

samples along the PC axes. In this book, **U** is referred to as the score matrix. (Note that in the literature both **U** and **US** are referred to as scores.) The **V** matrix contains the information about how the original measurements are related to the PCs (loadings). The **S** matrix is diagonal (has zeros everywhere except on the diagonal), and contains information about the amount of variance each PC describes.

4.2.2.1 PCA Example 1 The use of PCA will be demonstrated using a three-dimensional example. The discussion follows the "Six Habits of an Effective Chemometrican" which are detailed in Chapter 1.

Habit 1. Examine the Data

A plot of the data as intensity versus measurement (variable) number is shown in Figure 4.26 where each line represents a sample. This is the type of plot normally made to represent spectroscopic and chromatographic data. Look for any obvious outlier samples or questionable variables. With these data there do not appear to be any obvious outlier samples. That is, no samples appear to be different from the majority of the data. Also, there do not appear to be any particularly unusual variables, although with so few measurements the interpretation is limited. In cases where the data should form a smooth pattern when plotted in this way, it is sometimes possible to determine if a sample(s) or variable(s) is behaving very differently from the others. Any unusual samples and/or variables should be investigated.

Habit 2. Preprocess as Needed

If the measurements are known to be of different types with different measurement ranges, consider autoscaling. In all other situations when using PCA, the data should be mean centered because this allows the PCs to optimally describe the orientation of the samples in row space. Unless otherwise noted, all PCA discussions in this chapter assume mean-centered data.

To illustrate the importance of mean centering, PCA is performed on a matrix of data before and after mean centering. When the data are not mean centered (see Figure 4.27a), the first PC must describe the direction from the ori-

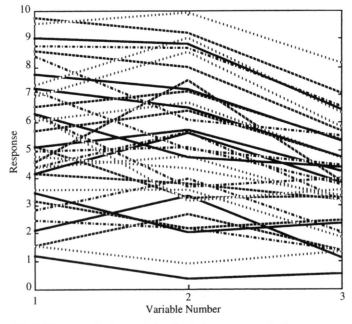

Figure 4.26. PCA example 1 raw data. Each line represents the measurements for a different sample.

gin to the cloud of data. The second PC is constrained to be orthogonal (per-pendicular) to the first and, therefore, also cannot orient along the length of the cloud. These two PCs span the space of the cloud, but are not oriented with the cloud in row space. Contrast this with the PCA analysis show in Figure 4.27b, where the PCs for the mean-centered data effectively describe the direction of the cloud of points. The first PC describes the length of the cloud

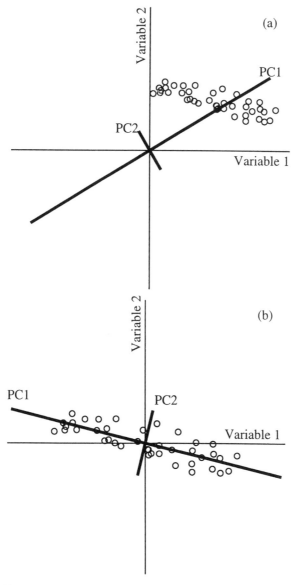

Figure 4.27. Example of PCA (a) without mean centering and (b) after mean centering.

and the second describes the width. After centering, the first PC is free to describe the main source of variation *within* the cloud of data.

Habit 3. Estimate the Model

The main parameters to define include the choice of preprocessing and, in some cases, the total number of PCs to estimate. The PCA outputs include the percent variance explained by each PC, cross validation results, scores, loadings, and residuals.

Habit 4. Examine the Results/Validate the Model

Several diagnostic tools are discussed below and a summary is found at the end of the section in Table 4.5. With PCA, these tools are used to investigate three aspects of the data set: the model, the samples, and the variables. The headings for each tool indicate the aspects that are studied with that tool. The primary use of the model diagnostic tools is to investigate the inherent dimensionality of the data set. The sample diagnostic tools are used to study the relationships between the samples and identify unusual samples. The variable diagnostic tools do the same, but for the variables.

Percent Variance Plot (Model Diagnostic): The percent of the total variation that each principal component axis describes can be used to determine the inherent dimensionality of the data set. This is accomplished by comparing the variation remaining using a given number of principal components to the expected noise level of the measurement variables. For example, assume that the error in the data comprises 10% of the variation in the data. In this case, keeping the number of PCs such that ~90% of the variation is described is appropriate.

Figure 4.28 shows the amount of variation described by the PCs calculated on the data in Figure 4.26 after mean centering. Nearly all the variation is described by the first two principal components, indicating that the samples lie on a plane in three-dimensional space. If the data were equally spread out in the two dimensions, the first two PCs would describe equal amounts of variance. Because the first factor describes much more variation than the second factor, it can be concluded that the data points form an object that is long and narrow.

Another important use of the percent variance is to temper the interpretation of loadings and scores plots. This is discussed below.

Measurement Residual Plot (Model, Sample, and Variable Diagnostic): The residuals are the portion of the original data that is not described by a given number of PCs. The number of elements in the residuals is the same as the number of measurement variables in the original matrix. To understand the residuals, consider applying PCA to a data set. When the first PC is estimated, it is possible to determine the variation that is described by this first dimension or factor. The contribution of this PC can be subtracted from the original data set, resulting in a residual for each sample. This process can be

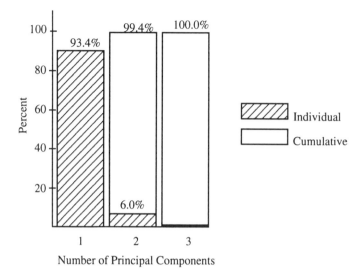

Figure 4.28. The percent variance explained by each principal component for PCA Example 1.

repeated, calculating subsequent PCs and subtracting their contributions. The result is a set of residuals for a given number of PCs. The residuals will resemble random noise when all of the relevant variation in a data set has been explained. The corresponding number of PCs can then be used as one indicator of the inherent dimensionality of the data.

In this example, the percent variation indicates that the data form a two-dimensional object (plane) in the measurement space. Figure 4.29*a* shows the residuals after removing the contribution from the first principal component. The magnitude and shape of the residual spectra are examined to determine if one PC is sufficient to describe the data. If the noise in the measurements is approximately ± 2 response units, one might conclude that one principal component is sufficient to describe the data. If the noise is known to be much smaller, more PCs should be used to describe the relevant variation. The structure in the residuals in Figure 4.29*a* is also an indication that additional PCs are required to describe the systematic variation in the data set. Figure 4.29*b* reveals considerably smaller residuals remaining after the contribution of the second PC is removed.

For this example, assume the noise in the data is small (approximately ± 0.5), and, therefore, two principal components are required. There still appears to be structure in the residual spectra, indicating that systematic variability remains. However, keep in mind that the objective is to describe the major sources of variation in the data set with the minimum number of dimensions. Some structure in the residuals is acceptable if this results in simplifying the problem into a small number of dimensions. (Because there are only three measurement

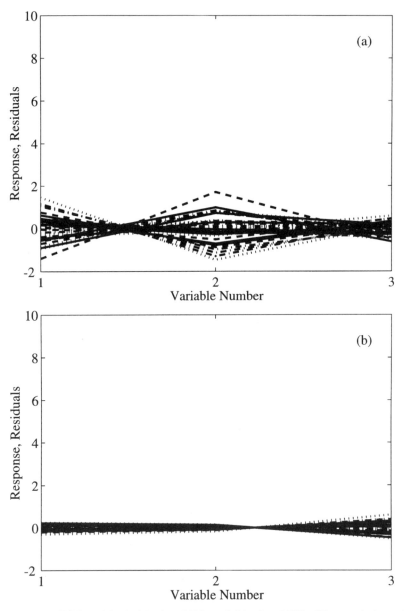

Figure 4.29. PCA residuals (*a*) after 1 PC and (*b*) after 2 PCs. The *y* axis is on the same scale as the original data.

variables, this data set can be exactly reproduced using three dimensions. There-fore, if a third PC is estimated, the residuals would be exactly zero.)

The residual plots can also be used to identify unusual samples. When one or more samples have a residual vector that is significantly different in magnitude or shape from the majority, they should be examined more closely.

There is no indication of unusual or outlying sample(s) in the residuals plotted in Figure 4.29. This indicates that the PCA model accounts for the systematic variation in all of the samples in the data set.

Root Mean Square Error of Cross Validation for PCA Plot (Model Diagnostic): As described above, the residuals from a standard PCA calculation indicate how the PCA model *fits* the samples that were used to construction the PCA model. Specifically, they are the portion of the sample vectors that is not described by the model. Cross-validation residuals are computed in a different manner. A subset of samples is removed from the data set and a PCA model is constructed. Then the residuals for the left out samples are calculated (cross-validation residuals). The subset of samples is returned to the data set and the process is repeated for different subsets of samples until each sample has been excluded from the data set one time. These cross-validation residuals are the portion of the left out sample vectors that is not described by the PCA model constructed from an independent sample set. In this sense they are like *prediction* residuals (vs. fit).

One advantage of the cross-validation residuals is that they are more sensitive to outliers. Because the left out samples do not influence the construction of the PCA models, unusual samples will have inflated residuals. The cross-validation PCA models are also less prone to modeling noise in the data and therefore the resulting residuals better reflect the inherent noise in the data set. The identification and removal of outliers and better estimation of noise can provide a more realistic estimate of the inherent dimensionality of a data set.

A common approach to cross-validation is called "leave-one-out" cross-validation. Here one sample is left out, a PC model with given number of factors is calculated using the remaining samples, and then the residual of the sample left out is computed. This is repeated for each sample and for models with 1 to n PCs. The result is a set of cross-validation residuals for a given number of PCs. The residuals as a function of the number of PCs can be examined graphically as discussed above to determine the inherent dimensionality. In practice, the cross-validation residuals are summarized into a single number termed the Root Mean Squared Error of Cross Validation for PCA (RMSECV_PCA), calculated as follows:

$$\text{RMSECV\_PCA}_k = \sqrt{\frac{\sum_{i=1}^{\text{nsamp}} \sum_{j=1}^{\text{nvars}} \left(\text{residual}_k(i,j)\right)^2}{\text{nsamp}}} \qquad (4.2)$$

In Equation 4.2, RMSECV_PCA$_k$ is a measure of the magnitude of the residuals for a PCA model using k principal components. The term residual$_k(i,j)$ is the residual for the ith sample and jth measurement variable, nsamp is the number of samples, and nvars is the number of variables.

The RMSECV_PCA values are plotted versus the number of PCs to determine the inherent dimensionality. Ideally, the RMSECV_PCA will decrease quickly and level off as the optimum number of principal components is approached. The RMSECV_PCA decreases quickly because the systematic variation is being described and thus the structure is being removed from the residuals. After the inherent dimensionality is reached, additional PCs will only describe noise and, therefore, no significant decrease in the RMSECV_PCA is realized. A number of statistical and empirical tools can be used to evaluate when the plot is statistically "leveling off" (Malinowski, 1991). These can help in determining the optimum, but may not yield conclusive results. Visual examination of the plots coupled with an understanding of the system can also aid in the determination of the inherent dimensionality.

For this example, the RMSECV_PCA plot is shown in Figure 4.30. Because there are only three variables, a maximum of three PCs can be calculated and they describe the space completely (RMSECV_PCA$_3$ = 0). In most situations, the number of PCs calculated is larger, which makes it easier to see the "leveling off" effect.

The model diagnostic tools suggest the inherent dimensionality of this data set is 2. With two principal components, the percent variance described is 99.4% and the residuals appear reasonably random and are small in magnitude. The RMSECV_PCA is not revealing, but also does not contradict this conclusion.

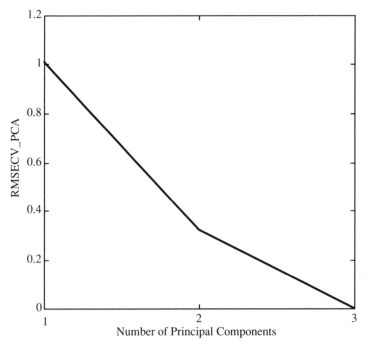

Figure 4.30. RMSECV_PCA from leave-one-out cross-validation for Example 1.

Loadings Plot (Model and Variable Diagnostic): Loadings are used to determine which variables are important for describing the variation in the original data set. The loadings are the cosine of the angle between the PC and the original variables and describe how the original measurement variables are related to each of the new PC axes. As the loading approaches 1 (or -1), the angle between the principal component and the original variable approaches 0 (or 180) degrees. This indicates the PC is parallel with the variable, meaning the variable contributes much variation to the PC. If the cosine is close to 0, the angle between the principal component and the original variable is close to 90 degrees. In this case, the variable is perpendicular to the PC and does not contribute to the variance described by that PC. It should be noted that only the absolute magnitude of the loading (without regard to sign) is discussed in this book. In some applications (Martens and Næs, 1989, pp. 132 and 138), the sign of the loadings is used for physical interpretations; here the primary use is in determining the relative contributions of the variables to the PCs.

It is important to note the amount of variation that is described by a PC when interpreting the loadings. A variable with a large loading value indicates that it contributes significantly to a particular PC. However, the variable may not be truly important if the PC does not describe a large amount of the variation in the data set.

Loadings can also be used to help determine the inherent dimensionality of the data set. This is accomplished by examining loading versus variable number plots and determining when the loadings begin to resemble random variation. This works well when the data come from instruments that yield measurement vectors with structured features (e.g., spectroscopy and chromatography). The relevant PCs describe systematic variation and, therefore, the loadings will also have structure. However, this approach is not appropriate when the measurement vectors do not have an expected, well-defined shape.

The example data set does not have well-defined features and, therefore, the loadings are not used to help determine the inherent dimensionality. For a demonstration of this use of loadings plots, refer to Section 4.3.2.1 where PCA is applied to data with structured features. For this example, the loadings for the first two principal components are shown in Figure 4.31. For PC1, the magnitude of the loadings is roughly equivalent, indicating that the variance (or spread of the data points) described by this PC is equally distributed amongst the three variables. The loadings for PC2 indicate that variables 1 and 2 have large loadings relative to variable 3. The loading for variable 3 is near zero, that is, the angle between PC2 and this variable is approximately 90 degrees. This means the second PC does not point in the same direction as the third variable.

The interpretation of the nearly equal loadings for PC1 is that the data do not preferentially align with any one of the original measurements directions. For PC2, the near zero loading for variable 3 indicates that the next dimension of the object lies roughly on the plane spanned by variables 1 and 2.

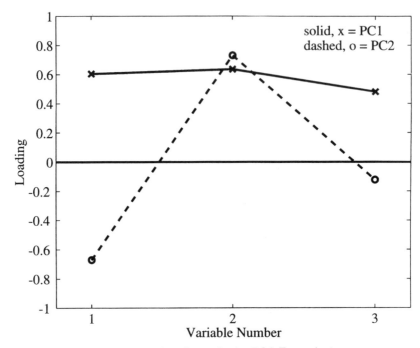

Figure 4.31. Loadings plot for PCA Example 1.

Another use of the loading plots is to identify unusual variables, as indicated by unexpected features in the plot. If the measurement system was such that the data vectors had discernible peaks, one would expect the loadings to contain features with approximately the same width as the original data features. In Figure 4.31 no unusual features are observed (however, owing to the nature of this data set, this diagnostic is not useful).

Scores Plot (Sample Diagnostic): The scores are the coordinates of the samples in the new coordinate system where the axes are defined by the principal components. These new axes are used to view the relevant variation in the data set in a smaller number of dimensions. The plot reveals how the samples are related to each other given the measurements that have been made. Samples that are close to each other on a given score plot are similar with respect to the original measurements provided the plot displays a sufficient amount of the total variation. This mathematical proximity translates to chemical similarity if meaningful measurements have been made.

As mentioned above, there are two representations of score plots, **US** and **U**. Both of these representations are prevalent in commercial software, and so it is important to note the type of scores that are generated when making interpretations. The choice of the **U** or **US** scores plot is a matter of personal preference or may be dictated by the software package. Figure 4.32*a* displays the **US** scores for PC2 versus PC1 for the example data. This two-dimensional

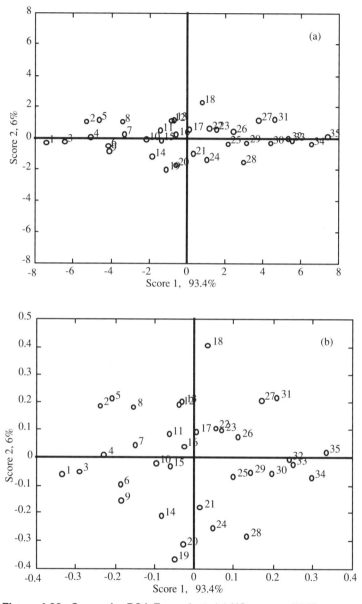

Figure 4.32. Scores for PCA Example 1: (a) **US** scores; (b) **U** scores.

plot accounts for 99.4% of the variation in the original data set, which is the sum of the variation described by the first two PCs. The object formed by the samples is long and narrow, as was also indicated by the percent variance explained. The negative score values are a result of mean centering the data. This plot captures a large percentage of the variation contained in the original data set; therefore, we can be confident that the relative distances between the data points are preserved. Comparisons about nearness of samples in the original row space can therefore be made based on this score plot. For example, it can be inferred that samples 1 and 3 are more similar than samples 1 and 35 with respect to the measurements.

The other score representation, U scores, is plotted in Figure 4.32b. This plot does not preserve the relative interpoint distances because the scores are normalized. For this reason, it is extremely important to note the amount of variation that is described by each PC when interpreting these plots. For example, although the distribution of the points in Figure 4.32b does not indicate a long and narrow object, this conclusion is drawn when the variation information is considered. That is, a small amount of variation is described by PC2 relative to PC1 (6.0% vs. 93.4%). The advantage of the U scores is that it is easier to see the relative distances between points in the direction of the PC describing less variance. The disadvantage is that it is easy to conclude that samples 18 and 24 are as dissimilar as samples 1 and 35 using Figure 4.32b. However, this conclusion is not drawn when the variance is considered.

Another skill that is useful to acquire when examining score plots is the ability to examine each dimension independently to determine which PC is accounting for any clustering. For example, PC1 is distinguishing between samples 5 and 31, while they appear to be similar when only PC2 is considered (i.e., the score values on PC1 are different for these samples but roughly equal on PC2). In contrast, PC2 is primarily responsible for discriminating the small differences between samples 18 and 24.

In other applications where more than two principal components are required to describe the relevant variation, multiple pairwise plots can be made. In this book, only selected pairwise plots are examined. If three-dimensional graphics are available, three dimensions can be viewed at one time.

Summary of Validation Diagnostic Tools for PCA: In conclusion, PCA has revealed the following about the example data set:

1. Over 99% of the object can be described using two dimensions.
2. The object formed is long (93.4%) and narrow (6.0%).
3. The scores plot reveals which samples are similar.
4. The long side of the object points equally in the direction of variables 1, 2, and 3.
5. The short side of the object is parallel to the plane formed by variables 1 and 2.

For this example, an examination of the complete three-dimensional row space (Figure 4.33) can be used to verify these conclusions. This plot shows the mean-centered data and the principal component axes. This plot is consistent with the conclusions from the PCA analysis and it is clear that the two PCs are effectively describing the object.

Table 4.5 summarizes the validation diagnostic tools discussed in this sec-

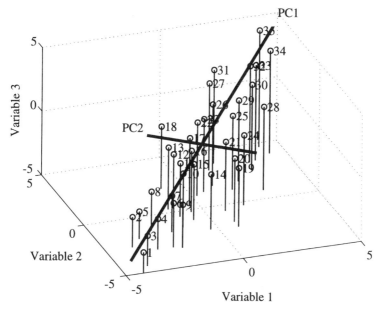

Figure 4.33. Mean-centered data for PCA Example 1, where the principal component axes are shown.

tion. The first column in the table lists the name of each tool and the second column describes results from both well-behaved and problematic data.

Habit 5. Use the Model for Prediction

Habit 6. Validate the Prediction
Habits 5 and 6 are not described because PCA is not used in this section as a predictive tool. The supervised pattern-recognition technique, SIMCA, uses PCA for class prediction and the details of Habits 5 and 6 for SIMCA are presented in Section 4.3.2.1.

4.2.2.2 PCA Example 2   Although the previous example was instructive, the true power of PCA is realized when examining data sets with larger numbers of variables. In this example, the data from a sensor array with 14 variables is discussed. In an industrial setting, gas-phase chemical information is needed for a number of reasons. These include industrial hygiene (e.g., room air quality) and process control (e.g., monitoring the concentration in a reactor headspace). Simple analytical sensors can be designed that fulfill many of the physical requirements, such as ruggedness and low cost, but they can suffer from poor selectivity. To overcome the selectivity shortcoming, multiple sensors with different

TABLE 4.5. Summary of Validation Diagnostics for PCA

Diagnostics	Description and Use
Model	
Percent variance plot (% variance explained vs. number of PCs)	Displays the fraction of the total variation in the data set that is described by the PCs.
	Used to determine the number of relevant PCs (i.e., inherent dimensionality) and interpret the diagnostics that follow.
	What to expect depends on the shape and dimensionality of the object formed by the samples in row space.
	With problematic data the inherent dimensionality that is indicated is not consistent with what is expected.
Measurement residual plot [$(\mathbf{r} - \hat{\mathbf{r}})$ vs. variable number]	The residuals are the portion of the sample measurement vector that is not explained using a given number of PCs.
	Used to determine the number of relevant PCs (i.e., inherent dimensionality).
	As the number of PCs used in the model approaches the inherent dimensionality the residuals decrease in magnitude and become more random.
Root mean square error of cross-validation for PCA plot (RMSECV_PCA vs. number of PCs)	Quantifies the magnitude of the measurement residuals using a given number of PCs.
	Used to determine the number of relevant PCs (i.e., inherent dimensionality).
	The curve always decreases as PCs are added. It levels off when the inherent dimensionality is reached.
Loadings plot (loadings vs. variable number)	For a given PC, the absolute magnitude of the loading for a measurement variable defines how that variable contributes to the PC. The patterns within the loading vectors from different PCs are examined.
	When the measurement vectors are continuous in nature (e.g., infrared spectra) the loadings can help determine the number of relevant PCs.
	A PC corresponding to a loading vector that resembles random variation is not considered relevant.
Sample	
Scores plot [PCy vs. PCx (vs. PCz)]	The scores represent how the samples are related to each other given the measurement variables. The plots are two- or three-dimensional representations of the row space.
	Used to examine the differences and similarities between the samples in the data set.
	Closeness of samples in the plot is interpreted as chemical similarity. The location of the samples is expected to be consistent with any a priori information.
	Be sure to note whether the scores are raw or normalized (**US** vs. **U**) and use the percent variance described by each PC when making interpretations.

TABLE 4.5. Continued

Diagnostics	Description and Use
	Often the primary information from PCA is obtained by examining the scores plots.
Measurement residual plot [$(\mathbf{r} - \hat{\mathbf{r}})$ vs. variable number]	The residuals are the portion of the sample measurement vector that is not explained using a given number of PCs.
	Used to identify outlying samples.
	An outlier is identified as a sample with a residual vector that is significantly different from the other samples.
	Examining the variables where a sample has unusual features can be used to determine the cause of the problem.
Variable	
Loadings plot (Loadings vs. variable number)	For a given PC, the absolute magnitude of the loading for a measurement variable defines how that variable contributes to the PC.
	Used to identify both important and unusual variables.
	A variable that contributes much to a PC has a large absolute loading (max = 1). Because the structure of the loading vector depends on the data being analyzed, the definition of an unusual variable is problem dependent.
	For well-behaved continuous data, the features of the loadings vectors are comparable to the preprocessed data.
Measurement residual plot [$(\mathbf{r} - \hat{\mathbf{r}})$ vs. variable number]	The residuals are the portion of the sample measure ment vector that is not explained using a given number of PCs.
	Used to identify unusual variables.
	Problematic variables are identified as those with unusual residual features for all of the samples.

response characteristics can be used to construct an array. Rather than a single sensor output, a measurement vector is generated for each sample. Principal components analysis is used to help determine if a given array has sufficient selectivity to distinguish between the components of interest. (Further analysis of this data set using a supervised pattern-recognition technique is presented in Section 4.3.2.2.)

Taguchi-type semiconductor sensors doped with various catalysts were used to form the array. Combustion of organic vapors takes place at the surface of the sensors and this reaction is influenced by the type and amount of various catalysts. Traditionally, these sensors are operated under steady-state conditions of the vapor flowing past the array. The steady-state sensor intensity reflects the amount of combustion. In this study, the sensors are used in a

TABLE 4.6. Chemicals Measured Using the Sensor Array

Class	Species
0	Air (background)
1	Acetone
2	Toluene
3	Cyclohexane
4	Methyl ethyl ketone
5	Ethyl acetate
6	Methanol
7	Hexane
8	1-Propanol
9	Benzene
10	Triethylamine

flow-injection mode and therefore temporal information is also obtained. The signal increases with time owing to (1) the diffusion of the sample in the chamber containing the sensors and (2) the kinetics of combustion at the surface of the sensors. For these experiments, the maximum slope of the signal in time was calculated in addition to the steady-state intensity. This resulted in a total of 14 measurements per sample (a slope and maximum intensity for each of seven sensors). The 10 pure chemical species (classes) listed in Table 4.6 are considered for classification. Samples with varying concentrations (0–3 ppt in air) were prepared, resulting in a total of 180 measurement vectors. (Hool et al., 1994)

Habit 1. Examine the Data

A plot of the 180 response patterns from the sensor array is shown in Figure 4.34a. The steady-state intensities for the seven sensors are indicated by i1–i7 and the sensor slopes as s1–s7. The most striking characteristic of these data is that the sensor intensities are much larger in magnitude than the slopes. Figure 4.34b is a graph of the sensor slopes using a different scale for the y axis. No samples or variables appear unusual in either of these plots. One indication that the sensors are functioning properly is that all seven sensor-intensity variables are of the same order of magnitude, as are all the sensor-slope variables. The most clear message from Figure 4.34 is that preprocessing is necessary to give the sensor intensities and slopes comparable weight in the calculations. This is not unexpected, because the units for the maximum intensity and slopes are different.

Habit 2. Preprocess as Needed

Autoscaling is performed in order to remove the scale difference between the sensor intensities and slopes. This involves subtracting the mean and dividing by the standard deviation for each column (see Chapter 3).

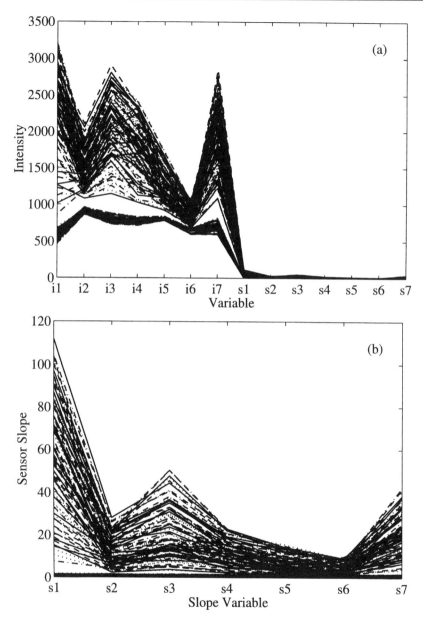

Figure 4.34. (*a*) Raw data for PCA Example 2; i1–i7 are the intensities for sensors 1–7, and s1–s7 are the sensor slopes. (*b*) Slope data only.

Habit 3. Estimate the Model

The data are autoscaled and the following outputs for four PCs are calculated: the percent variance explained by each PC, scores, loadings, and residuals.

Habit 4. Examine the Results/Validate the Model

Percent Variance Plot (Model Diagnostic): The percent variance described by each factor shown in Figure 4.35 reveals that more than 92% of the variance is described by the first principal component. Without further investigation it is not clear whether other principal components are also significant.

Measurement Residual Plot (Model, Sample, and Variable Diagnostic): Figure 4.36a–c shows the residuals after using 1, 3, and 4 principal components, respectively. The residuals have been converted back from the autoscaled units to the original measurement units to facilitate comparisons. In this example, these plots do not definitively indicate the inherent dimensionality of the data set.

Root Mean Square Error of Cross Validation for PCA Plot (Model Diagnostic): The leave-one-out cross-validation results are shown in Figure 4.37. The RMSECV_PCA appears to decrease smoothly, but again the selection of the inherent dimensionality is not obvious. As is sometimes the case, none of the model diagnostic tools has definitively indicated the dimensionality of the data set. However, much of the variability (99.1%) is described by the first four PCs and the remaining diagnostics are used to further the understanding of the structure of the data.

Loadings Plot (Model and Variable Diagnostic): The first 3 loadings are plotted in Figure 4.38. There are no variables with loadings close to ±1, revealing that no PC is closely aligned with any one sensor measurement. The loadings for principal component 1 are approximately the same for all variables. This is because the measurements were autoscaled, and so no conclusions can

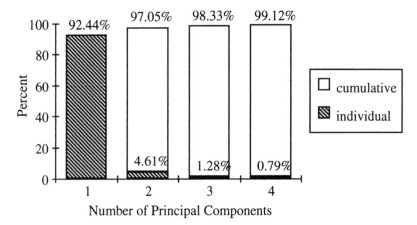

Figure 4.35. The percent variance explained by each principal component for PCA Example 2.

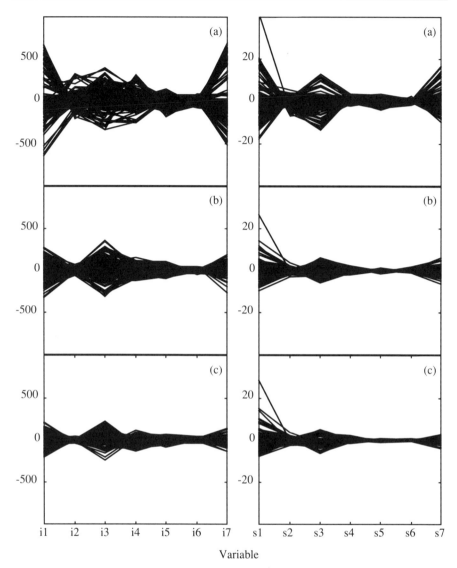

Figure 4.36. Residuals of the PCA Example 2 data after including (*a*) one, (*b*) three, and (*c*) four principal components. The residuals have been converted back to the original measurement units.

be made about the relative sensitivity of individual sensor measurements. For PC2, the intensity from sensor 7 and the slope from sensors 5 and 7 contribute most significantly. The intensities on sensors 2 and 6 and the slope on sensor 5 contributes most to PC3. This information is useful when designing a sensor array for a specific application (Carey et al., 1986).

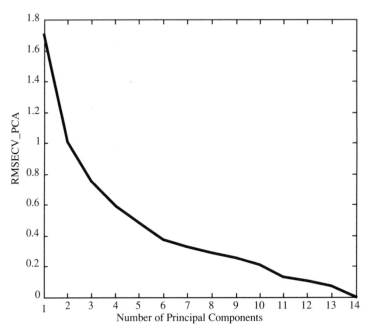

Figure 4.37. RMSECV_PCA from leave-one-out cross-validation for PCA Example 2.

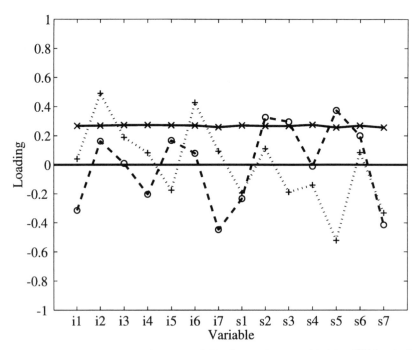

Figure 4.38. The first 3 loadings for PCA Example 2: solid (x) = PC1; dashed (O) = PC2; and dotted (+) = PC3.

The loadings plot is also examined for inherent dimensionality and unusual variables. No conclusions can be drawn because of the limited utility of this diagnostic when examining data sets comprised of discrete measurement variables.

Scores Plot (Sample Diagnostic): Figure 4.39*a* shows the plane spanned by principal components 1 and 2, representing 97% of the variation in the data. The axis labels display the amount of variation that each PC describes. There is a tight cluster of points from which two lines of points extend. PC1 discriminates the tight cluster from the lines while PC2 primarily distinguishes the

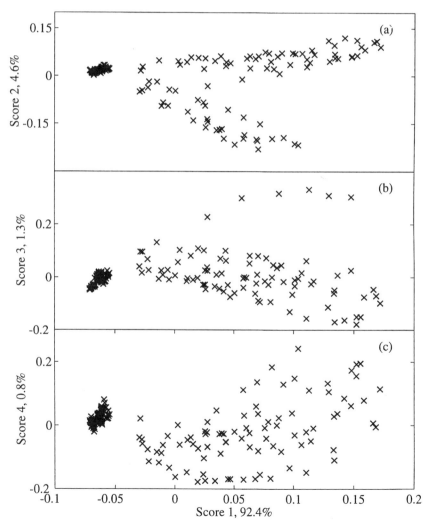

Figure 4.39. Scores for PCA Example 2: (*a*) PC2 vs. PC1; (*b*) PC3 vs. PC1; and (*c*) PC4 vs. PC1.

lines from each other. Figure 4.39*b* displays the scores for PC3 versus PC1, showing a total of 94% of the variation in the data. Again there is a tight cluster of points, but now with a thick band of points and a curved line containing six points. PC3 clearly differentiates the six points in the upper line from the rest of the data. Figure 4.39*c* shows PC4 versus PC1, representing 93% of the variation in the data. The clustering in this graph is limited, but there is one tightly clustered collection of points and a larger cloud of data. In this last plot, PC1 is responsible for all of the clustering. From these scores plots and the percent variation described, we conclude that the data lie primarily in a three-dimensional space described by principal components 1–3. The data form a collection of tightly clustered samples with three fingers. This can be seen more clearly by examining a three-dimensional scores plot (Figure 4.40).

More in-depth understanding of the object formed by the data can be obtained given additional knowledge of the data (e.g., concentration and chemical identity of the samples). Figure 4.41 shows Figures 4.39*a* and *b* with concentration information. This plot reveals that the tightly clustered samples are the background measurements that were taken before each injection (concentration = 0). The size of the cluster indicates that the noise in the blank is considerably smaller than the variation owing to concentration and chemical differences. A second feature observed in these plots is that the concentration increases when moving along the three fingers away from the background cluster, indicating that PC1 is primarily describing concentration variation.

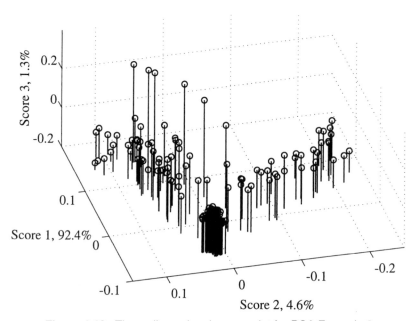

Figure 4.40. Three-dimensional scores plot for PCA Example 2.

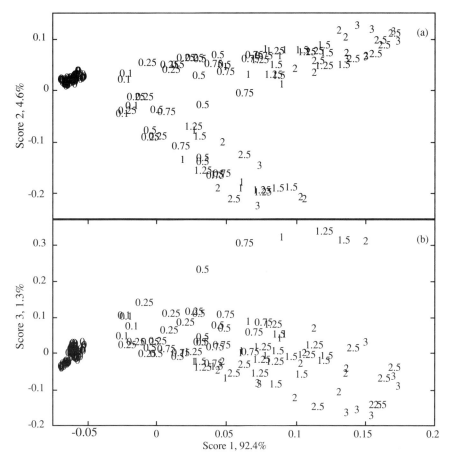

Figure 4.41. Scores for PCA Example 2: (a) PC2 vs. PC1; (b) PC3 vs. PC1. The numbers are the concentrations (ppt) of the organic solvent in the sample.

Figure 4.42 shows the same scores plots now with a number indicating the chemical identity of each sample (see Table 4.6). In Figure 4.42a the lower finger contains classes 2 (toluene), 3 (cyclohexane), 7 (hexane), and 9 (benzene). The upper finger contains classes 1 (acetone), 4 (methyl ethyl ketone), 5 (ethyl acetate), 6 (methanol), and 8 (1-propanol). Figure 4.42b shows that the lone finger protruding above the PC2 vs. PC1 plane (in the direction of PC3) contains class 10 (triethylamine). Examining the chemical nature of the clusters reveals that the lower finger in Figure 4.42a contains components with only carbon and hydrogen. The upper finger contains components that contain carbon, hydrogen, oxygen, and nitrogen. Finally, the third finger seen in Figure 4.42b contains the only compound in the set comprised of carbon, hydrogen, and nitrogen. These plots demonstrate the ability of this array of sensors to discriminate between chemical functional groups.

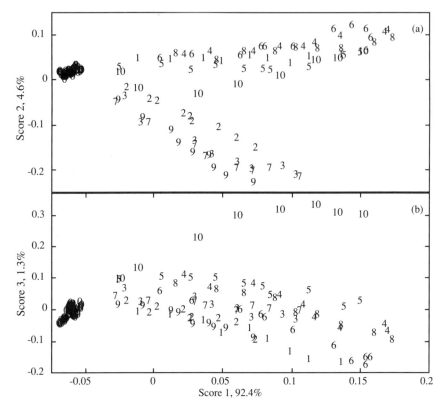

Figure 4.42. Scores for PCA Example 2: (*a*) PC2 vs. PC1; (*b*) PC3 vs. PC1. The numbers on the graph correspond to the chemical species (Table 4.6).

Summary of Validation Diagnostic Tools for PCA: In this example, the feasibility of using an array of sensors to distinguish between chemical compounds was investigated with PCA. The PCA results have demonstrated that the sensors are able to distinguish between functional groups. This is a positive result even though the 10 chemical classes were not separated into individual clusters in the scores plots. To develop a model to predict class membership of all of the chemical classes, one would need to use a supervised pattern-recognition technique. (See Section 4.3.2.2 where SIMCA is applied to this same data set.) The supervised methods use the class information in the calculations enabling them to more effectively distinguish between the classes.

Habit 5. Use the Model for Prediction

Habit 6. Validate the Prediction
Habits 5 and 6 are not described because PCA is not used for prediction.

4.2.3 Summary of Unsupervised Pattern Recognition

The goal of unsupervised techniques is to identify and display natural groupings in the data without imposing any prior class membership. Even when the
ultimate goal of the project is to develop a supervised pattern recognition
model, we recommend the use of unsupervised techniques to provide an initial view of the data.

Summary of HCA

The HCA technique examines the interpoint distances between the samples in
a data set and represents that information in the form of a two-dimensional
plot called a dendrogram. The HCA method is an excellent tool for preliminary
data analysis. It is useful for examining data sets for expected or unexpected
clusters, including the presence of outliers. It is informative to examine the
dendrogram in conjunction with PCA because they give similar information in
different forms.

Weaknesses:

1. Unlike PCA, HCA does not indicate which variable(s) contribute most to
the clustering. It is therefore not possible to evaluate the discriminating power
of the different variables.

2. Not all of the geometric information is retained in the dendrogram. This
can make the interpretation of poorly resolved clusters difficult.

3. There is only one view of the data which is presented—the dendrogram.
There is no interactive way of manipulating the dendrogram to allow the user
to explore the data using human pattern-recognition capabilities.

Strengths:

1. One advantage of HCA over PCA is that the dendrogram represents all of
the variation in the original data set. This is in contrast to PCA, which typically
only presents some fraction of the total variation in the scores plots.

2. Unlike PCA, with HCA there is no need to determine the rank of the data
matrix. This is especially useful for looking at a snapshot view of the data
when the inherent dimensionality of the data set exceeds three.

3. Having only one view of the data (dendrogram) is a weakness, but is also
a strength. There are some advantages to a simple standardized presentation.

Summary of PCA

Principal components analysis is used to obtain a lower dimensional graphical
representation which describes a majority of the variation in a data set. With
PCA, a new set of axes are defined in which to plot the samples. They are constructed so that a maximum amount of variation is described with a minimum
number of axes. Because it reduces the dimensions required to visualize
the data, PCA is a powerful method for studying multidimensional data sets.

Like HCA, it is an excellent tool for preliminary data exploration. It is useful for examining data sets for expected or unexpected clusters, including the presence of outliers.

Weaknesses:

1. The inherent dimensionality (rank) of the data set must be determined. This process is not always straightforward.

2. Although the number of dimensions is decreased, it is still possible to become inundated with plots. The utility of PCA is limited with data sets with high inherent dimensionality because multiple pairwise or three-dimensional plots must be used to visualize the data.

Strengths:

1. PCA displays multidimensional row spaces with a few well-chosen plots. It does a good job of presenting the data in the majority of applications.

2. PCA can be used to filter noise from data sets.

3. It provides information about the measurement variables. The loadings indicate the amount of variation contained by each measurement variable.

4. Interactive tools can be used to explore the data given the PCA results. For example spinning three-dimensional plots of various combinations of scores can be examined.

Our recommendation is to use both of the methods when they are available. HCA gives a broad view of the data and PCA allows for a closer examination of samples and clusters that are highlighted in HCA.

4.3 SUPERVISED LEARNING

Supervised methods are used when the goal is to construct a model to be used to classify future samples. This is accomplished using a set of data with known classifications to "train" the computer to distinguish between classes. An example of a supervised model is one developed for classifying different raw materials. Measurements are taken on samples for each raw material (class) and a model is derived that best distinguishes between the classes. The class of incoming raw materials is predicted with the model to verify their identity and perhaps integrity before they are used in the chemical manufacturing process. This is possible if the analytical measurements are of sufficient sensitivity and selectivity to detect differences between raw materials.

Like the unsupervised methods, the supervised methods discussed in this book are based on the assumption that samples that are chemically or physically similar will be near each other in measurement (row) space.

4.3.1 K-Nearest Neighbor (KNN)

With KNN, the predicted class of an unknown sample is assigned as the class of the sample(s) nearest to it in multidimensional space. In this book, Euclidean distances are used to measure the nearness between samples in row space. The Euclidean distance between samples x and y is calculated in nvars dimensions as

$$\text{Distance} = \sqrt{(x_1 - y_1)^2 + (x_2 - y_2)^2 + \cdots + (x_{\text{nvars}} - y_{\text{nvars}})^2} \quad (4.3)$$

where x_i and y_i are the coordinates of samples x and y in the ith dimension of the row space (where i ranges from 1 to nvars). Other distance measures can be used [e.g., the Mahalanobis distance emphasizes some measurements over others (Massart et al., 1988)].

To classify an unknown, the distance is calculated between it and a set of samples with known class membership (training set). The closest K samples are then used to make the classification. To select the optimal K (the number of neighbors to poll for future classifications), a cross-validation procedure is applied to a set of data with known class identities. Each sample in the training set is treated as an unknown and is classified using the remaining training set samples. This is repeated using different numbers of nearest neighbors (K) for the classification. If the classes are well separated, K = 1 provides a good classification rule (i.e., the one nearest neighbor has a high likelihood of belonging to the same class). However, for more confidence in the classification, more nearest neighbors should be considered. Thus, K can be thought of as being the number of "votes" that are tallied to decide on the membership of the unknown. The more samples that agree on a particular classification, the more confidence that can be placed in the results.

An example of KNN classification is shown in Figure 4.43 where there are two measurements (variables 1 and 2) and four classes (A, B, C, and D). Several training samples are shown for each class along with three unknown samples (labeled X, Y, and Z). To classify X using a one-nearest-neighbor rule, one would simply find the one sample that is closest to X and assign that class to X. From Figure 4.43 it clear that X would be classified as an A using this approach. In fact, the first 10 nearest neighbors to X all belong to class A.

Note that using an 11 nearest-neighbor rule to classify X would result in 10 A votes and 1 D vote. The only reason a D sample receives a vote is because there are only 10 A samples. This does not pose a problem in this example because of the large number of A samples. However, if there were only 1 A sample and 10 D samples, an 11 nearest-neighbor classification wou
votes and 1 A vote even if the unknown is very close to the A
far from the D class. Therefore, the maximum value of K that s
sidered is equal to the number of samples in the class with th
bers.

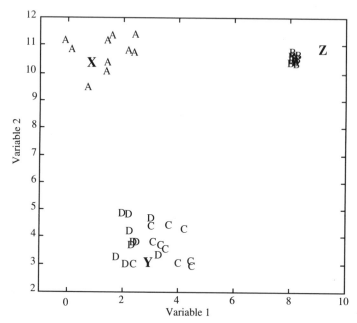

Figure 4.43. Training set samples in class A–D are displayed in the two-measurement row space. Unknown samples are labeled X, Y, and Z.

Unknown sample Y is contained within the overlapped classes C and D, and the classification is less obvious. The first nearest neighbor is from class D, while the second nearest is from class C. In the situation where multiple classes are represented in the K-nearest neighbors, most software packages classify an unknown into the class containing the majority of the nearest neighbors. For example, sample Y would be classified as a C using a three nearest-neighbor rule because two of the three nearest neighbors are from class C (see Figure 4.43). If there is a tie (e.g., K = 2, and both neighbors are from different classes), one method of resolution is to select the class of the nearest sample. Rules to resolve this conflict can be developed, but it may be the case that there is not enough discrimination between the classes given the set of measurements.

While assigning a class is the goal of KNN, it is also of interest to determine the confidence to place on the classification. Several approaches can be taken to measure this confidence. For example, more confidence is placed on classifications when all K nearest neighbors are from the same class. Conversely, the confidence in the classification decreases as the K nearest neighbors are represented by more than one class (e.g., the first nearest neighbor is from class A and the second nearest neighbor is from B).

A quantitative approach to placing a confidence on classification is to calculate the proportion of the K nearest neighbors that are members of the respective classes. In the terminology used in this book, this is an attempt to validate the prediction of class membership. For example, using K = 5 for classifying unknown sample X from Figure 4.43, all five nearest neighbors are from class A. Therefore 100% (5/5) of the nearest neighbors are from A. For the classification of unknown sample Y, 60% (3/5) of the nearest neighbors are from class C and 40% (2/5) of the nearest neighbors are from class D. Given the low percentage for the "most likely" class C, the prediction of this unknown as C would be considered questionable.

One situation where this approach would not flag a poor quality classification is illustrated by considering unknown Z in Figure 4.43. This sample does not appear to be a member of class B because it is not contained in the cluster comprised of B samples. However, with a 10 nearest-neighbor classification, the prediction would be 100% B. Given this large K, a correspondingly high confidence would be given to this classification, in spite of the fact that Z does not appear to belong to class B.

A better approach to validating the prediction is to compare the distance from the unknown sample to the predicted class relative to an "expected" distance for known members of that class. From the last example, the distance from Z to its nearest neighbor in class B is much larger than the distances between the samples within that class. This can be flagged by calculating a measure of expected interpoint distances for samples in each class. These distances are then compared to the distance of the unknown to the different classes to validate class membership. One algorithmic approach is discussed below illustrating the classification of unknown Z with respect to class B.

First calculate the distance (d_{unk}) from the unknown to the nearest neighbor in the class that was identified by KNN. (See Figure 4.44 for an expanded view of class B and unknown Z.) Next, calculate the interpoint one nearest-neighbor distances for each of the samples in the training set for the class being considered. These nearest-neighbor distances are shown by the solid lines between the Bs in Figure 4.44. The mean and standard deviation of these distances [$\bar{d}_B$ and sd(d_B)] are calculated and a "goodness value," G, is computed as follows:

$$G = \frac{d_{unk} - \bar{d}_B}{\text{sd}(d_B)} \tag{4.4}$$

This value is similar to a t value in statistics in that it indicates the number of standard deviation units the distance of the unknown is from the average class distance. The smaller the value of G, the more confidence one would have that the unknown belongs to that class. In fact, a negative goodness value is obtained if the distance of the unknown to the closest training sample (d_{unk}) is less than the average of the minimum interpoint distances ($\bar{d}_B$). Conversely, large G values are obtained when the distance of the unknown is large relative to the class interpoint distances. In practice, a positive cutoff value is set and unknown samples are excluded from the class if the calculated goodness value is larger than the cutoff value. The training set data are used to select an acceptable value for the cutoff.

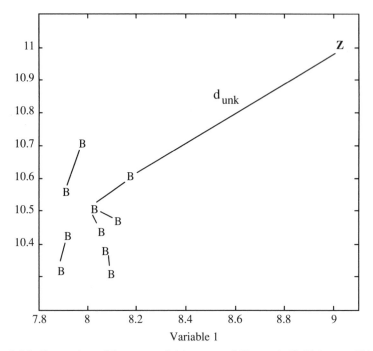

Figure 4.44. Expansion of the upper right corner of Figure 4.43. Unknown Z is classified by KNN as a member of class B, even though it is some distance away from the training set B samples.

4.3.1.1 KNN Example 1 The use of KNN is demonstrated using the data shown in Figure 4.43. The discussion follows the "Six Habits of an Effective Chemometrican" which are detailed in Chapter 1. These data are also discussed in Example 1 of HCA found in Section 4.2.1.1.

Habit 1. Examine the Data

The plot of the raw data is shown in Figure 4.45. There are 10 samples in each class and therefore K will be less than or equal to 10. When examining this graph, keep an eye out for unusual samples (there do not appear to be any here), unusual variables (difficult to determine with only two measurements), and the need for preprocessing. From this graph it appears as if classes A and B are well separated while classes C and D are overlapped.

Habit 2. Preprocess as Needed

From Figure 4.45 there does not appear to be any obvious preprocessing that is necessary. Also, there is no prior information which would suggest a need to preprocess.

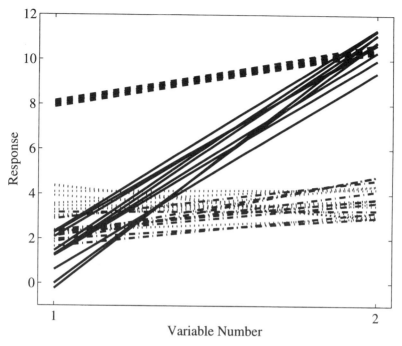

Figure 4.45. Raw data of the training set samples for Example 1. Class A = solid; class B = dashed; class C = dotted; class D = dashed–dotted.

Habit 3. Estimate the Model

To develop a KNN model, a distance measure is selected and the optimal number of nearest neighbors is determined. It is recommended that K be selected using leave-one-out cross-validation applied to a training set. The outputs of the analysis are the predicted classes for the training set and the goodness values.

Habit 4. Examine the Results/Validate the Model

Several diagnostic tools are discussed below and a summary is found at the end of the section in Table 4.9. With HCA these tools are used to investigate two aspects of the data set: the model and the samples. The headings for each tool indicate the aspects that are studied with that tool. The primary uses of the model diagnostic tools are to choose an appropriate value for K and to assess how well the classes are separated. The sample diagnostic tools are used to more closely investigate the clustering and identify unusual samples.

Incorrect Classification Plot (Model Diagnostic): The first diagnostic to examine is a plot of incorrect classifications versus the number of nearest neighbors, shown in Figure 4.46.

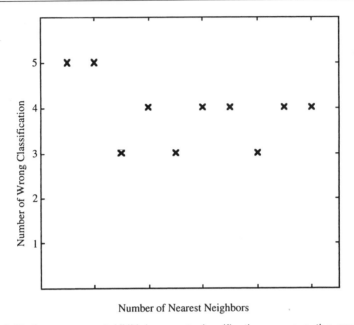

Figure 4.46. Leave-one-out KNN incorrect classifications versus the number of nearest neighbors for Example 1.

When one or two nearest neighbors (K = 1 or 2) are used there are five samples incorrectly classified. The minimum number of misclassifications is achieved using K = 3, 5, or 8 with each producing three misclassified samples. These values of K are all good candidates for the final model.

Classification Table (Model Diagnostic): The classification table can aid in determining which classes are not well separated. It summarizes how the samples from each of the classes are classified for a given value of K. Table 4.7 is the Classification Table for this example when considering three nearest neighbors. The first row indicates that the 10 samples in class A are all classified as belonging to class A. Similarly, the second row indicates that all 10 class

TABLE 4.7. Classification Table with Three Nearest Neighbors for KNN Example 1

Classified As → Known Class ↓	A	B	C	D
A	10	0	0	0
B	0	10	0	0
C	0	0	9	1
D	0	0	2	8

B samples are also correctly classified. Rows three and four show that one class C sample was predicted to be in class D, and two class D samples were predicted to be in class C. This is consistent with the results in Figure 4.46, which indicate that three samples were misclassified when using three nearest neighbors. The table indicates that with three nearest neighbors, classes C and D overlap.

Remember that a sample would be classified as belonging to class A if two out of three of its nearest neighbors are from class A. Therefore, Table 4.7 does not indicate small overlap between the classes. The next diagnostic tool is useful for more closely examining the clustering between classes.

Nearest-Neighbor Table (Model and Sample Diagnostic): The nearest-neighbor table is a more detailed summary of the classification than the previous two diagnostics. For each sample, leave-one-out cross-validation is used to predict the K nearest neighbors and these are displayed along with the true class. Table 4.8 shows the first, second and third nearest neighbors for each of the samples in the example data set and reveals that all three nearest neighbors for each sample in classes A and B are from the same (correct) class. This provides confidence that classes A and B are well resolved from the other classes in the set. On the other hand, samples from classes C and D do not have consensus between all the nearest neighbors. For example, of the 10 C samples, six had at least one of its nearest neighbors from class D and one sample was misclassified (sample 28). This is an indication that classes C and D are not well resolved.

This table can also be examined to identify possible outliers. If, for example, sample 1 was labeled as an A, but all of its closest neighbors were from class B, a labeling error is suspected. In Table 4.8 there are misclassifications of samples in the C and D classes, but the large number of errors is more an indication of overlap than outliers. It is possible that there multiple outliers, but outliers cease to be outliers when they are the majority.

From these diagnostics, it has been determined that classes C and D are not completely separated and therefore there will be some ambiguity with the classification of future samples in these two classes. If this is not acceptable, different measurements can be investigated to better separate classes C and D. Another approach is to consider these two classes as one class (CD). Repeating the calculation with only three classes (A, B, and CD) results in no misclassifications. For illustration purposes, it will be assumed that the overlap of these classes is acceptable.

Goodness Value Plot (Model and Sample Diagnostic): In prediction, a unanimous classification does not guarantee that an unknown is close to the samples in the predicted class even if the classes were found to be well separated (refer back to Figure 4.44). Therefore, the "goodness" value in Equation 4.4 is used to evaluate the quality of the classification using a relative distance measure. The approach for validating the prediction is to evaluate the distance of the unknown to the predicted class relative to an internal measure of how diffuse the samples are in that class.

TABLE 4.8. Nearest-Neighbor Table for the Three Nearest Neighbors for KNN Example

Sample No.	True Class	Class of First Nearest Neighbor	Class of Second Nearest Neighbor	Class of Third Nearest Neighbor
1	A	A	A	A
2	A	A	A	A
3	A	A	A	A
4	A	A	A	A
5	A	A	A	A
6	A	A	A	A
7	A	A	A	A
8	A	A	A	A
9	A	A	A	A
10	A	A	A	A
11	B	B	B	B
12	B	B	B	B
13	B	B	B	B
14	B	B	B	B
15	B	B	B	B
16	B	B	B	B
17	B	B	B	B
18	B	B	B	B
19	B	B	B	B
20	B	B	B	B
21	C	C	C	D
22	C	C	C	C
23	C	D	C	C
24	C	C	D	C
25	C	C	C	C
26	C	C	C	C
27	C	C	C	C
28[a]	C	D	D	D
29	C	C	D	C
30	C	C	C	D
31	D	D	D	D
32[a]	D	C	C	D
33	D	D	D	D
34	D	D	D	D
35	D	C	D	D
36	D	D	D	D
37[a]	D	C	C	C
38	D	D	C	D
39	D	D	D	D
40	D	D	D	D

[a]Misclassified by majority vote.

To determine a cutoff for the goodness value, the training set data are used in a leave-one-out scheme. The goodness values are calculated for every training set sample for the class predicted by the K nearest neighbors. For example, a goodness value for sample 1 is calculated for class A (see Table 4.8). These leave-one-out goodness values (see Figure 4.47 for the K = 3 results) are then used to select a cutoff for separating acceptable and unacceptable classifications. Four is a reasonable cutoff to use because the largest goodness value is slightly less than this value. An unknown sample that yields a goodness value of less than 4 is assumed to be a valid classification. A goodness value greater than 4 indicates a poor classification.

The goodness value can also be used to flag possible outliers within the training set. If one sample has an unusually large goodness value, further investigation is warranted. What is deemed unusual is a function of the problem at hand. The key is to look at the goodness values obtained from the training set and make a qualitative assessment of the acceptability of the largest goodness value. For example, the goodness value for sample 8 in Figure 4.47 does not appear to be unusual given the other goodness values. The conclusion is therefore that no outliers are present in the training set.

PCA and HCA (Sample Diagnostic): In addition to the KNN numerical results, it is instructive to use other pattern-recognition tools to examine the data. As discussed in Sections 4.2.1 and 4.2.2, HCA and/or PCA can be helpful for visualizing multivariate data to better understand the clustering (see Section 4.2.1.1 for HCA results using a superset of these data).

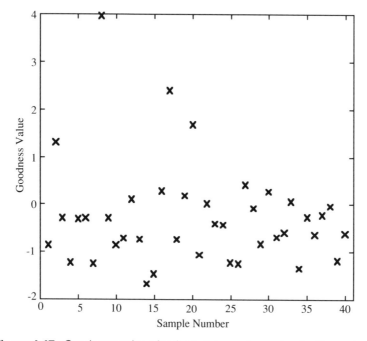

Figure 4.47. Goodness values for the training set samples for Example 1.

Summary of Validation Diagnostic Tools for KNN: Based on the diagnostic tools, 3 is an appropriate value for K when the classification validation for unknown samples is performed using the goodness values. If the validation is to be done by considering what percentage of the near neighbors belong to each class, a larger K should be investigated (e.g., K = 8).

Table 4.9 summarizes the validation diagnostic tools discussed in this section. The first column in the table lists the name of each tool and the second column describes results from both well-behaved and problematic data.

Habit 5. Use the Model for Prediction

Prior to prediction it is important to preprocess unknown samples using the same methods as were applied to the training set samples. Next, the distances from the unknown to all members of the training set are calculated. Then the K closest samples are polled for a classification based on a majority vote (using K determined in the model validation). If there is a tie, the predicted class is selected based on the first nearest neighbor.

The three nearest-neighbor model developed above is used to classify four unknown samples. Based on majority vote, the predicted classes are: unknown 1 = D, unknown 2 = A, unknown 3 = D, and unknown 4 = A.

Habit 6. Validate the Prediction

Several prediction diagnostic tools are discussed below and a summary is found at the end of the section in Table 4.12. These tools are used to assess the reliability of the classification results.

Nearest-Neighbor Table: The nearest-neighbor table for the four unknowns shown in Table 4.10 indicates that unknown samples 1 and 3 do not have reliable classifications because the votes are not unanimous. The fact that samples from classes A and D are amongst the three nearest neighbors for unknown sample 1 is of concern because the training set samples for classes A and D are well resolved. Because classes C and D are overlapped in the training set, the result for unknown sample 3 is not surprising. In contrast, the classification of unknowns 2 and 4 can be considered reliable because all nearest neighbors are from the same class.

Goodness Value: The predicted class and goodness values for the four unknowns are shown in Table 4.11 and indicate that the classification of unknown 1 is suspect because it has a value greater than 4. This is consistent with the nearest-neighbor table result for this sample where the vote is not unanimous. Unknown 2 has an acceptable goodness value, which is consistent with the unanimous vote for class A for this sample. Unknown sample 3 has an acceptable goodness value even though the votes from the nearest neighbors are not unanimous. This is not surprising because classes D and C are known to be overlapped. Finally, unknown 4 has a large goodness value even though it received a unanimous vote for class A. These samples are further investigated using raw data plots.

TABLE 4.9. Summary of Validation Diagnostics for KNN

Diagnostics	Description and Use
Model	
Incorrect classification plot (number of incorrect classifications vs. K)	Shows the number of misclassifications using different values for K (K = number of nearest neighbors).
	Used to help select an optimal K. Indicates which values of K warrant further investigation.
	For well-behaved data, the number of misclassifications will be acceptable for some value(s) of K.
	With problematic data the number of misclassifications is too large for all values of K.
Classification table (table of actual vs. predicted class for a given K)	For a given class (row in the table), shows the number of samples predicted to be members of the classes represented in the training set. A table can be generated for any value of K.
	Used to help select an optimal K and indicates the classes that are overlapped.
	For ideal data, all the samples are classified correctly and the table only has numbers on the diagonal.
	When the classes are not distinct, many samples are incorrectly classified and the table has off-diagonal entries.
Nearest-neighbor table (table of K nearest neighbors for all samples)	Lists the classes of the K nearest neighbors for each sample in the training set.
	Used to help select an optimal K and indicates the classes that are overlapped.
	For ideal data all of the K nearest neighbors for the training set samples belong to the correct known class.
Goodness value plot (G vs. sample number)	G is a quantitative measure of the quality of a classification.
	Used to determine a cutoff value for use in the prediction phase.
Sample	
Nearest-neighbor table (table of K nearest neighbors for all samples)	Lists the classes of the K nearest neighbors for each sample in the training set.
	Used to identify problematic samples.
	A problematic sample has one or more K nearest neighbors that belong to an incorrect class(es).
HCA, PCA	Additional tools for visualizing the relationship between the samples (see Sections 4.2.1 and 4.2.2).
Goodness value plot (G vs. sample number)	G is a quantitative measure of the quality of a classification.
	Used to identify problematic samples.
	Unusual samples have large goodness values relative to the other samples in the training set.
Variable	No variable diagnostics are discussed.

TABLE 4.10. Nearest-Neighbor Table for the Prediction of the Four Unknowns

Unknown	First Nearest Neighbor	Second Nearest Neighbor	Third Nearest Neighbor
1	A	D	D
2	A	A	A
3	D	C	D
4	A	A	A

Raw Measurement Plot: In Figure 4.48 the raw data for the four unknown samples are plotted with the mean of the training samples for the four classes (A-D). The mean is the average of the measurement vectors for all samples within a class and is used as a representation of the expected features for samples that belong to the respective class. The differences in the features may indicate the source of problems for unknowns where the classifications are suspect. Figure 4.48 confirms the other diagnostic tools, which indicate that: (*a*) unknown 1 is not a member of any class, (*b*) unknown 2 is a member of class A, (*c*) unknown 3 is a member of class D, and (*d*) unknown 4 is most like A but has different features.

Because there are only two variables in this example, it is also possible to plot the entire row space (see Figure 4.49). Unknown samples 2 and 3 lie well within the predicted classes, as indicated by the goodness values in Table 4.11. In contrast, unknowns 1 and 4 are not within any cluster and have correspondingly high goodness values. Figure 4.49 shows why unknown 4 was classified unanimously as belonging to class A. This illustrates how the goodness value calculation augments the information from the nearest-neighbor table when vaiidating predictions.

Summary of Prediction Diagnostic Tools for KNN: Based on the diagnostics, the conclusion is that unknown sample 2 is a member of class A and unknown sample 3 is a member of either class C or D. The classifications of unknown samples 1 and 4 are considered unreliable.

TABLE 4.11. Goodness Values for the Four Unknown Samples

Unknown	Predicted Class	Goodness Value
1	D	13.0
2	A	1.0
3	D	0.1
4	A	12.7

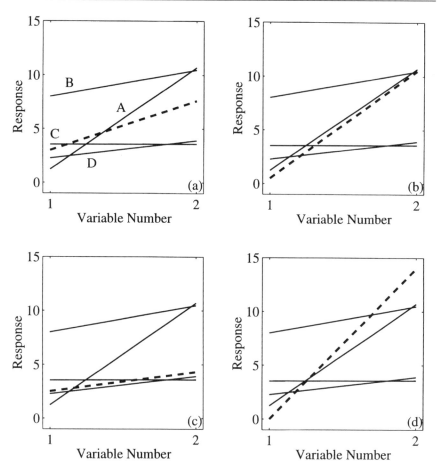

Figure 4.48. The four unknowns (dashed lines) and the average of the training set classes A–D. (*a*) Unknown 1; (*b*) unknown 2; (*c*) unknown 3; (*d*) unknown 4.

Table 4.12 summarizes the prediction diagnostic tools discussed in this section. The first column in the table lists the name of each tool and the second column describes results for both well-behaved and problematic data.

4.3.1.2 KNN Example 2 The goal of the project described in this example is to determine whether NIR reflectance spectroscopy can be used for sorting recycled plastic containers. The KNN classification is selected because it is simple and does not make assumptions about the statistical distribution of the classes. This is important because the number of samples for most of the classes in the training set is small (3–5) and, therefore, it is not possible to check any distributional assumptions.

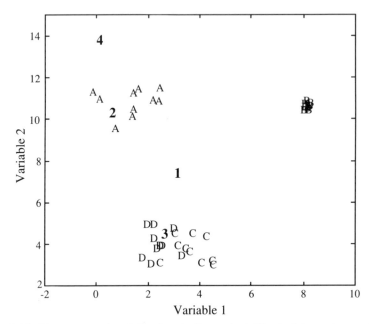

Figure 4.49. Row space plot of the data in Example 1. The training set samples are labeled according to the class, and the unknowns are labeled 1–4.

TABLE 4.12. Summary of Prediction Diagnostics for KNN

Prediction Diagnostics	Description and Use
Nearest-neighbor table	A list of the class(es) of the K nearest neighbors for a given prediction sample.
	Used to help assess the reliability of the classification of unknown samples.
	To have confidence in a classification, nearly all of the K nearest neighbors should be from the same class.
	A sample classification is suspect if the K nearest neighbors are from multiple classes.
Goodness value	The goodness value is a quantitative measure of the quality of a classification.
	Used to help assess the reliability of the classification of unknown samples.
	To have confidence in a classification, the goodness value must be less than the cutoff determined in the model-validation phase.
Raw Measurement Plot (**r** vs. variable number)	A plot of the measurement vector for the prediction sample.
	Used in conjunction with knowledge of the chemistry and the measurement system to assign cause when suspect classifications are encountered.

TABLE 4.13. Samples Used for the Waste Plastic
Sorting NIR Feasibility Study

Plastic Type	Number of Samples
Polyethylene terephthalate (PET)	13
Polypropylene (PP)	3
Polystyrene (PS)	5
Polyvinyl chloride (PVC)	4
Polyethylene (PE)	4

The following waste plastic samples (from used containers) were gathered and the reflectance spectra were measured in the range 1100–2500 nm (see Table 4.13).

Habit 1. Examine the Data

The spectra for the 29 training set samples are shown in Figure 4.50 (the baseline is corrected at 1600 nm and the classes are offset for clarity). Ideally, the spectra for each sample within a class would overlay and the features would be different between the classes. In Figure 4.50, there appears to be significant within-class variation, which is addressed in the next section. No unusual samples are observed, but this finding is reevaluated after preprocessing is applied.

Habit 2. Preprocess as Needed

For the preprocessing of these data, there are physically meaningful reasons for each of the steps. These coincide with the main objective of making the spectra of the samples within a class look similar (i.e., reducing the within-class variability). The large within-class variation is due primarily to the difference in scatter inherent in reflectance measurements. This effect can be reduced by performing a second derivative (see Figure 4.51). Even after the second derivative is taken, large within-class variations remain. However, by selecting the 1600- to 1800-nm region, these variations are reduced (see Figure 4.52). This wavelength region encompasses the first overtones of the CH_3, CH_2, CH, and ϕ-CH transitions. It is reasonable to expect that this region can be used to distinguish between these polymers. Another source of variation, pathlength, can be removed by normalizing the spectra (see Section 3.1.1), as shown in Figure 4.53.

Although the within-class variations have been reduced significantly, one or two samples in each class are still somewhat different from the others in the class. However, they appear to be members of their respective classes (i.e., the shapes of the spectra are approximately the same). Furthermore, because there are so few samples in most classes, it is not reasonable at this point to exclude any sample from the analysis.

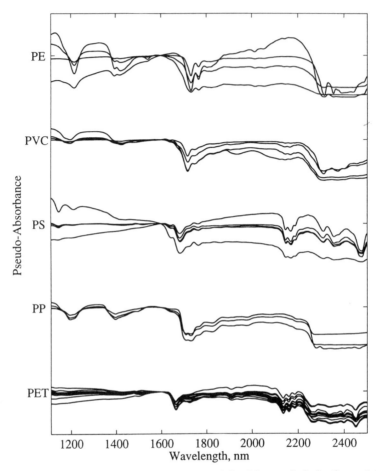

Figure 4.50. Near-infrared reflectance spectra of the 29 recycled plastic containers, baseline corrected at 1600 nm.

Habit 3. Estimate the Model

The preprocessed data and class membership information is submitted to the analysis software. Euclidean distance and leave-one-out cross-validation is used to determine the value for K and the cutoff for *G*.

Habit 4. Examine the Results/Validate the Model

Incorrect Classification Plot (Model Diagnostic): The results from the leave-one-out KNN classification as a function of K are shown in Figure 4.54. The calculations were performed for K = 1–3 because the class with the smallest number of samples (PP) has three members. For any of these values of K, the classifications are always correct for all 29 samples. This implies that the classes are all well separated.

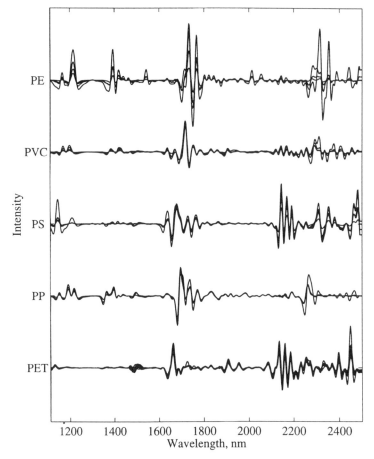

Figure 4.51. Near-infrared reflectance spectra of the 29 recycled plastic containers, preprocessed with a second derivative.

Classification Table (Model Diagnostic): The classification table also indicates that the classes are clearly distinguishable using these measurements (see Table 4.14). This table is identical whether K = 1, 2, or 3.

Nearest-Neighbor Table (Model and Sample Diagnostic): The nearest-neighbor table (not shown) for K = 3 indicates that in nearly all cases the three nearest neighbors are from the same class. The only exception is for the three polypropylene samples where the third nearest neighbor is from class PE. Because there are three PP samples, when one is left out in cross-validation, only two PP samples remain in the training set. Therefore, even with distinct classes, one of the three nearest neighbors must be from a different class. This does not necessarily indicate a problem with overlapping classes, just a limitation in the number of samples in the training set. When predictions are performed, all three of the PP samples are used in the training set.

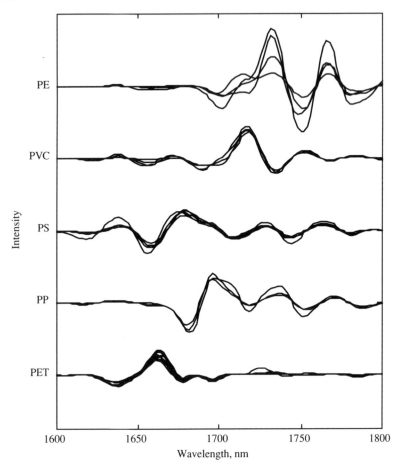

Figure 4.52. Near-infrared reflectance spectra of the 29 recycled plastic containers, preprocessed with a second derivative and variable selection.

Goodness Value Plot (Model and Sample Diagnostic): A goodness value is calculated for all samples in the training set (Figure 4.55) except PP. The PP samples do not have goodness values because it is not possible to calculate a leave-one-out goodness value with only three samples in a class. (When one sample is left out of the class, there are only two remaining samples, the standard deviation term in Equation 4.4 goes to 0, and the goodness value approaches infinity.) A few other samples had large negative goodness values and are not displayed.

The largest goodness value corresponds to a PS sample and it appears to be unusual when compared to the rest of the training set samples (all other goodness values are less than 4). This unusual spectrum can be distinguished in Figure 4.53, where the other four PS spectra are tightly clustered together rela-

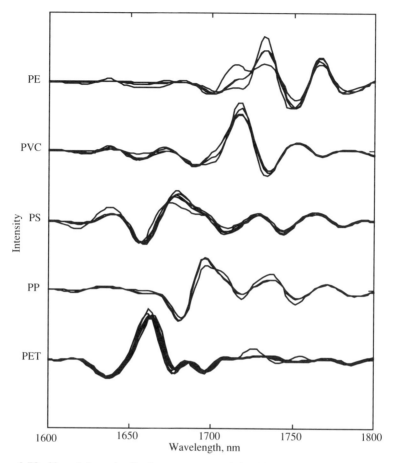

Figure 4.53. Near-infrared reflectance spectra of the 29 recycled plastic containers, preprocessed with a second derivative, variable selection, and normalization.

tive to this sample. In Figure 4.53, the other classes also appear to have unusual samples, but the goodness values are not large because the other spectra in the class are not as tightly clustered. Eliminating the unusual PS sample can be considered, but with so few samples in the PS class it is difficult to conclusively identify outliers. The approach taken is therefore to perform predictions and use a cutoff of 8 for the goodness value. If more samples are added to the training set, this cutoff should be reevaluated.

PCA and HCA (Sample Diagnostic): A superset of these data is discussed in Section 4.2.1.2.

Summary of Validation Diagnostic Tools for KNN: The conclusion is that the NIR approach for sorting the five plastics examined in this study worked well using K = 3 and a cutoff for *G* = 8.

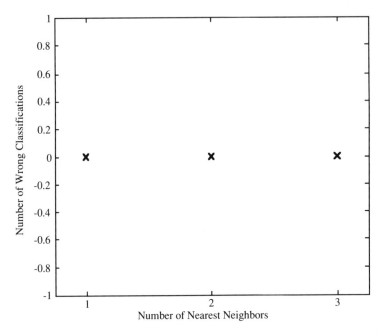

Figure 4.54. Leave-one-out KNN incorrect classifications versus the number of nearest neighbors for Example 2.

Habit 5. Use the Model for Prediction

Three unknown samples are examined in the prediction phase. The results of a three nearest-neighbor classification are: unknown 1 = PET; unknown 2 = PET; unknown 3 = PVC.

Habit 6. Validate the Prediction

Table 4.15 lists the nearest-neighbor table and goodness values for the unknown samples.

Nearest-Neighbor Table: The classes of the three nearest neighbors are shown in columns 2–4 of Table 4.15. For unknowns 1 and 2, all three nearest

TABLE 4.14. Classification Table with One, Two, or Three Nearest Neighbors

Predicted Class → Known Class ↓	PET	PP	PS	PVC	PE
PET	13	0	0	0	0
PP	0	3	0	0	0
PS	0	0	5	0	0
PVC	0	0	0	4	0
PE	0	0	0	0	4

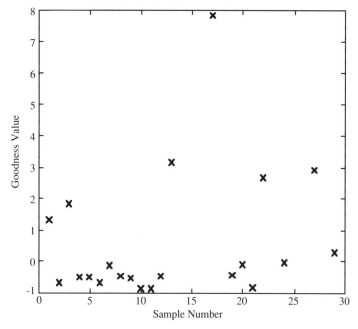

Figure 4.55. Goodness values for the training set samples for Example 2.

neighbors are from the same class. In Habit 4, all classes appear to be well separated and, therefore, the unanimous vote for these unknowns implies the classification is reliable. Unknown 3, on the other hand, has two neighbors which are PVC and one neighbor which is PET. This classification is not reliable and is investigated further.

Goodness Value: The goodness values for the three unknowns are listed in the last column of Table 4.15. Consistent with the nearest-neighbor table, unknowns 1 and 2 have a goodness value less than 8, and so are considered reliable classifications. Unknown 3 has a goodness value greater than 8, further indicating a problem with the classification.

TABLE 4.15. Nearest-Neighbor Table and Goodness Values of Three Unknown Samples

Unknown	First Nearest Neighbor	Second Nearest Neighbor	Third Nearest Neighbor	Goodness Value
1	PET	PET	PET	−0.7
2	PET	PET	PET	1.0
3	PVC	PVC	PET	13.8

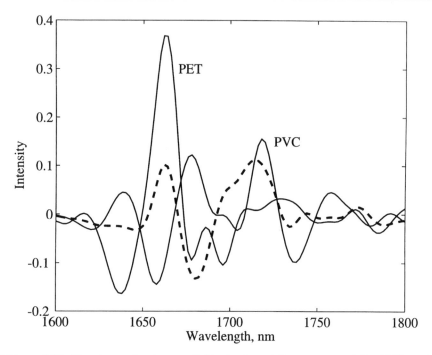

Figure 4.56. Spectrum of unknown 3 (dashed) and the average of the PVC and PET training set spectra.

Raw Measurement Plot: If an on-line analyzer were sorting the containers, unknown 3 would have been rejected and the spectrum stored for further evaluation. The spectrum of this unknown is plotted in Figure 4.56 along with the average of the training set spectra for PVC and PET. Unknown 3 has features from both of these classes and, therefore, the classification results are not surprising.

Summary of Prediction Diagnostic Tools for KNN: In this example, it has been demonstrated that NIR coupled with KNN can distinguishing between different materials used in plastic containers. A simple prediction diagnostic tool was also presented and applied to three unknown samples. Two of the three unknowns were classified with high confidence.

4.3.2 SIMCA (Soft Independent Modeling of Class Analogies)

SIMCA uses PCA to model the shape and position of the object formed by the samples in row space for class definition. A multidimensional box is constructed for each class and the classification of future samples (prediction) is performed by determining within which box, if any, the sample lies. This is in contrast to KNN, where only the physical closeness of samples in space is used for classification.

To construct the multidimensional boxes, a training set of samples with known class identity is obtained. The training set is divided into separate sets, one for each class, and principal components are calculated separately for each of the classes. The number of relevant principal components (rank) is determined for each class and the SIMCA models are completed by defining boundary regions for each of the PCA models.

A plot of the three-dimensional row space of samples from three classes is shown in Figure 4.57. The class A samples occupy a line, class B samples occupy a plane, and class C samples occupy a three-dimensional space. This concept can be extended to higher-dimensional objects where "hypercubes" are drawn around the data points.

To predict the class of an unknown, it is necessary to determine what region of measurement space it occupies. Mathematically, this is accomplished by projecting the measurement vector of the unknown sample into each of the SIMCA models. A graphical illustration is shown in Figure 4.57, where unknown X would be classified as a member of class A and not as a member of the other two classes. In prediction then, a sample may be a member of one, several, or no classes. If it is a member of more than one class, this is because the measurement system and the SIMCA models do not have enough discrimination power to distinguish between classes. If the unknown is not a member of any of the classes in the training set, this is an indication that the sample is unusual relative to the training set (it can be considered an outlier). This may be due to an error in measurement, or the sample may be the result of some unusual or unknown chemistry.

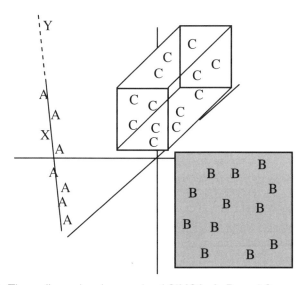

Figure 4.57. Three-dimensional example of SIMCA. A, B, and C represent samples from three different classes, and the X and Y samples are unknowns.

Recall from Section 4.2.2 that PCA merely defines a coordinate system in which to represent samples. For SIMCA, boundaries are also placed on the axes defined by this coordinate system. Without these boundaries, unknown Y in Figure 4.57 would be classified as a member of class A even though it is far from the cluster of the class A samples. The boundary for the SIMCA model for class A is defined as the ends of the solid line passing through the class A samples. Using this boundary avoids the misclassification of unknown Y.

SIMCA models can be conceptualized as distinct boxes containing the training set samples for each class (see Figure 4.57). The PCA representation defines the orientation of the box in measurement space and the boundaries define the edge of the box. To determine whether an unknown sample is within a class, SIMCA uses a combination of the distance of the unknown from the PCA model and the distance from the boundary. This is illustrated in Figure 4.58 for the prediction of unknown sample Y using class A data modeled with one principal component. The distance c is the distance from unknown Y to the PCA model (i.e., c is the PCA residual). The distance b is the distance from the projection of Y onto the principal component model (Y_{PCA}) to the SIMCA box boundary.

The value that is used to determine the closeness of Y to class A is a function of a^2 (where $a^2 = b^2 + c^2$). The a^2 is converted to a variance and divided by the variance observed for the class A samples to form an F value, called F_{calc}. A critical value is chosen empirically or from an F table and is compared with F_{calc}. The unknown sample is classified as an A if F_{calc} is less than the critical value. Other methods of expressing this concept (e.g., p values) are based on the same statistical criterion. The user should refer to the software manual if the procedure encountered differs from that discussed in this section.

Although the box drawn around the class has straight edges, the region in space that defines whether an unknown sample is included in a class is curved because of the nature of the F test. For example, the acceptance region for a one-dimensional class is shaped like a hot dog. Similarly, the acceptance region for a two-dimensional model resembles a compressed ice cream sand-

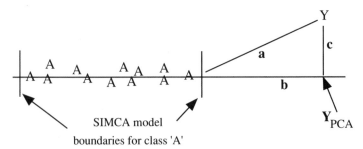

Figure 4.58. Prediction of unknown Y with one-dimensional class A. Distance c corresponds to the PCA residual. Distance b is the distance from the class boundary to the projection of Y onto the principal component (point Y_{PCA}).

wich with the ice cream bulging out. In all cases, the size and curvature of the acceptance region is related to the number of samples in the calibration set and the critical value that is selected for prediction (parameters in the F test). To simplify the drawings, we show only straight-edged boxes around the class with an understanding that the shape of the true acceptance regions are slightly different.

Once the class boundaries are defined, it is important to determine whether any of the classes in the training set overlap. This indicates the discriminating power of the SIMCA models and will impact the confidence that can be placed on future predictions. There are various algorithmic measures of class overlap and the reader is referred to their software package documentation for details. In this chapter, class overlap is indicated when training set samples are predicted to be members of multiple classes. This is demonstrated in a two-dimensional example shown in Figure 4.59. Two classes are shown where class A is described by one principal component and class B is described by two principal components. The overlap of the classes is indicated because unknown Z is classified as belonging to both classes.

The training set samples may be predicted to be in none, one, or multiple classes. It is tempting to think that if there is overlap of classes A and B, that all class A samples will be predicted to be in classes A and B, and vice versa. However, this is not the case. Figure 4.60 shows some examples of overlapped classes in two dimensions. Only samples in the intersection of two classes are predicted to be in both classes. In Figure 4.60a and c, the extent of overlap involves only a fraction of the total samples in both classes. Class B in Figure 4.60b is entirely contained in another class and therefore all the class B samples are predicted to be in both classes A and B.

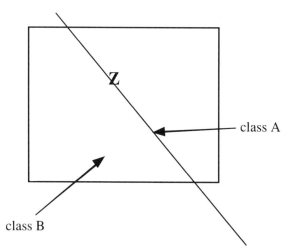

Figure 4.59. Class A (line) and Class B (box) overlap. Sample Z is an unknown.

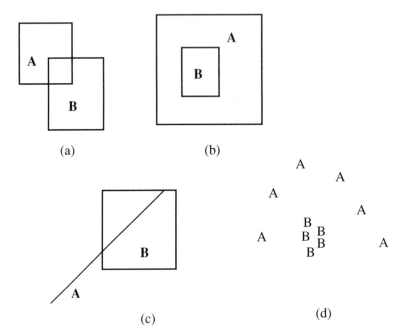

Figure 4.60. Four examples of overlapped classes. (*a*) and (*c*) Partially overlapped classes; (*b*) one class contained within another; (*d*) an example of class orientation that is difficult to model using SIMCA or PCA.

Figure 4.60*d* shows one limitation of the SIMCA approach where PCA uses two principal components to fit class A. The acceptance region is a two-dimensional box that overlaps the region occupied by the class B samples. Class B is also described using a smaller two-dimensional box. The result is a situation similar to that as shown in Figure 4.60*b* where in this case class B samples are classified as being in both classes. Both of these cases can be identified using the diagnostic tools discussed below.

4.3.2.1 SIMCA Example 1   The details for constructing a SIMCA model are described in this section following the "Six Habits of an Effective Chemometrician," which are described in Chapter 1. The data set discussed in this example consists of three classes with 27 samples in each class. The samples for classes A and B were obtained through natural sampling of a system (no formal design was used), while the class C samples were collected using a formal experimental design (three-level factorial of three components). It is not a requirement that each class in a training set have the same number of samples. The primary requirement is to have enough samples in each class to cover the space occupied by future samples.

Habit 1. Examine the Data

A plot of the measurement vectors of all of the samples in the training set is shown in Figure 4.61 (an offset for each class was added for clarity). The measurement vectors within a class have similar features and there is significant overlap of features between classes. Examine this plot for outlier samples and/or measurement variables as well as an indication of the need for preprocessing. In this case, all measurements appear to be reasonable and preprocessing does not appear to be warranted.

Two samples in class B have a much larger response than the other samples in the class, although the general shape is the same as the rest of the class members. Because the samples for this class were not collected according to a

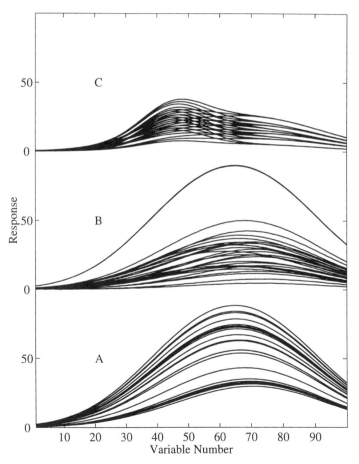

Figure 4.61. Library data for the three classes in SIMCA Example 1. The offset was added for clarity.

formal experimental design, these samples may simply be extreme points. On the other hand, these samples may have been mishandled or mislabeled during data collection. At this point, it is not possible to determine the root cause for these extreme points. At a minimum, it should be verified that the correct classification labels have been used for the training set.

Habit 2. Preprocess as Needed
The data are mean centered for each of the classes independently before building the SIMCA models. No other preprocessing was considered necessary.

Habit 3. Estimate the Model
A software package is used to calculate the SIMCA model. The steps we use to generate the SIMCA models are found in Table 4.16.

Steps 1 and 2 are discussed in detail in Sections 4.2.1 and 4.2.2 (PCA and HCA). In step 3, the training set is divided into calibration and test sets to facilitate the estimation of the SIMCA models. Typically, we leave more than half of the data in the calibration set. It is also a good practice to repeat the calibration procedure in Table 4.16 with different selections of calibration and test sets. An alternative to separate test sets is to implement some form of cross-validation. For example, leave-one-out cross-validation can be performed where each sample is left out and predicted one at a time.

Once the calibration and test sets are defined, SIMCA models are estimated using initial rank estimates (from PCA) and default class volumes (from the software package). The samples in the test set are predicted and the results evaluated to determine if modification of the ranks and/or class volumes is needed (see Habit 4 for details of this evaluation). If the number of misclassifications is unacceptable, the rank and class volume parameters are adjusted and the process repeated.

TABLE 4.16. Steps for Constructing SIMCA Models

1. Calculate principal components for the samples in the individual classes and determine the initial setting for the rank of each class model.
2. Perform unsupervised pattern recognition on the entire training set to see if the classes appear overlapped (PCA and/or HCA).
3. Divide each class in the training set into calibration and test sets.
4. Using the calibration set, construct SIMCA models for each class with initial settings for rank and boundary distance.
5. Predict the class of the test set samples using the initial SIMCA models.
6. Evaluate test set prediction results.
7. If the classification results are not acceptable, return to step 4 with different settings. If classification is acceptable, go to step 8.
8. Using the combined calibration and test sets, construct final SIMCA models for each class using the rank and boundary settings determined in steps 4–6.

Habit 4. Examine the Results/Validate the Model

Different diagnostic tools are used to examine first the PCA then the SIMCA output. A summary of the PCA diagnostics that are applied here is found in Table 4.5 in Section 4.2.2.1. These tools are used to investigate three aspects of the data set: the model, the samples, and the variables. The headings that follow for each tool indicate the aspects that are studied with that tool. The primary use of the model diagnostic tools is to determine the rank of each class. The sample diagnostic tools are primarily used to identify unusual samples with variable diagnostic tools doing the same, but for the variables. After examining the results from PCA and selecting the rank for each class, the SIMCA model(s) are constructed and validated.

In the following example only the details of the PCA analysis for class B will be examined. Nothing unusual was found from the separate PCA analyses of classes A and C. The rank for classes A and C were estimated to be one and three, respectively.

PCA of Class B—Percent Variance Plot (Model Diagnostic): The first principal component describes 99.15% of the variance, the second describes 0.85%, and the third describes less than 0.01%. Assuming the noise in the data is measured to be greater than or equal to 0.01% of the variation, one would infer that these data lie on a two-dimensional plane.

Measurement Residual Plot (Model, Sample, and Variable Diagnostic): The spectral residuals after one and two principal components are shown in Figure 4.62. After one principal component (Figure 4.62a), the residuals are small, but structured. After two principal components (Figure 4.62b), the residuals are smaller and more random in nature, further confirming that the data lie on a two-dimensional plane.

Root Mean Square Error of Cross-Validation for PCA Plot (Model Diagnostic): Figure 4.63 displays the RMSECV_PCA vs. number of principal components for the class B data from a leave-one-out cross-validation calculation. The RMSECV_PCA quickly drops and levels off at two principal components, consistent with the choice of a rank two model.

Loadings Plot (Model and Variable Diagnostic): The loading plot in Figure 4.64 reveals that the first and second loadings have nonrandom features, while the third is random in nature. This suggests a two-principal component model consistent with the percent variance explained, residuals plots, and RMSECV_PCA results

Scores Plot (Sample Diagnostic): The positions of the samples on the two-dimensional PCA plane are shown in the scores plot (Figure 4.65). A few samples to the right side of the graph are not clustered with the rest of the samples. These correspond to the samples with the highest responses for class B in Figure 4.61. The spectral residuals in Figure 4.62b are not unusual, indicating that these samples lie in the same plane as the rest of the samples in this class.

PCA on Entire Training Set: To examine the extent of class overlap, the next step is to perform PCA on the entire training set of samples. This could

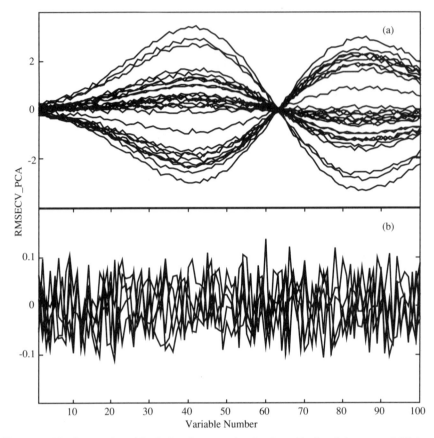

Figure 4.62. Spectral residuals for the samples in class B after (a) one and (b) two PCs.

have been performed first, but it is useful to have some feel for the rank of the individual classes before taking this step. It is also possible to use the information learned by examination of the individual classes to understand the entire data set.

The results from a PCA analysis of the entire training set can indicate the probability of success in using SIMCA. Ideally, each class will occupy a portion of space separate from the other classes. If the entire data set occupies a space with rank larger than 3, it will only be possible to see slices of the space when using two- or three-dimensional plots. Overlap observed in these two- or three-dimensional slices may not be significant because the separation may actually be taking place in one of the unobserved higher dimensions.

The rank of the PCA model for the entire training set is at least equal to 3 because this is the largest rank from any of the individual classes. The rank will equal 3 if the other classes are contained within the same space. The largest rank possible is the sum of the ranks of the individual classes assuming the

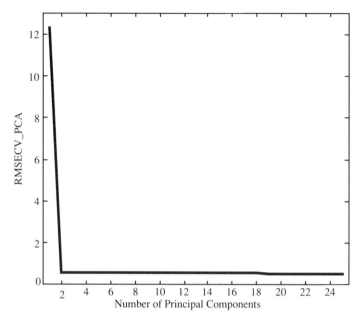

Figure 4.63. RMSECV_PCA from leave-one-out cross-validation of class B.

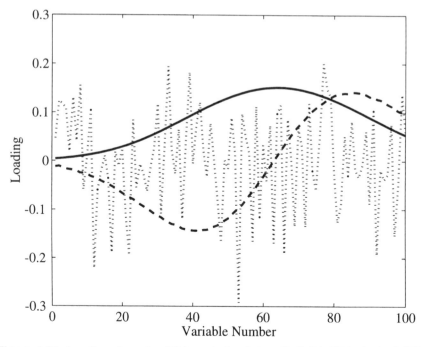

Figure 4.64. Loadings from the PCA analysis of class B. Solid, PC1; dashed, PC2; dotted, PC3.

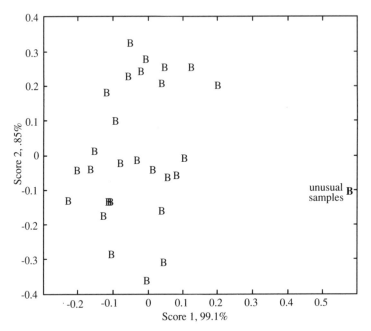

Figure 4.65. Score 2 versus score 1 for the class B PCA.

classes do not occupy any redundant regions of space. In this example, the maximum rank is $1 + 2 + 3 = 6$.

Percent Variance Plot (Model Diagnostic): The percent variances described for the entire training set are shown in Table 4.17. These results indicate that the three classes occupy three or four dimensions in row space. Further examination of the other diagnostics will help refine this estimate.

Measurement Residual Plot (Model, Sample, and Variable Diagnostic): The residuals plots (not shown) after two principal components showed structure, but after three principal components no significant structure remained.

TABLE 4.17. Percent Variance Described by the PCA of the Entire Training Set

PC No.	Variance (%)
1	97.37
2	2.57
3	6.2e–02
4	4.7e–05
5	3.8e–05
6	3.6e–05
7	3.3e–05

Root Mean Square Error of Cross Validation for PCA Plot (Model Diagnostic): The RMSECV_PCA vs. number of principal components for a leave-one-out cross-validation displayed in Figure 4.66 indicates a rank of 3.

Loadings Plot (Model and Variable Diagnostic): Loadings 1–4 are displayed in Figure 4.67. The first three have a definite nonrandom behavior compared with the fourth. There is a limited amount of structure in loading four, indicating a model of size 3 or 4 factors.

Scores Plot (Sample Diagnostic): The scores plots can now be used to examine the class overlap in the three-dimensional space that describes the majority of the data set. For a three-dimensional space it is possible to construct three two-dimensional score plots (score 2 vs. score 1, score 3 vs. score 1, and score 3 vs. score 2). For this example only the two most informative score plots are shown (see Figure 4.68). In the score 2 vs. score 1 plot, classes A and C are not overlapped and, therefore, samples from these two classes should easily be distinguished using SIMCA. Class B overlaps with both classes A and C and there are some unusual class B samples that are far from the bulk of the other class B samples. These unusual class B samples are the same samples that appeared unusual in the PCA of class B (Figure 4.65). Because they lie on the same line with the class A samples, it might be concluded that they actually belong to class A but have been mislabeled. Because of the overlap of the class B samples with the other two classes, it

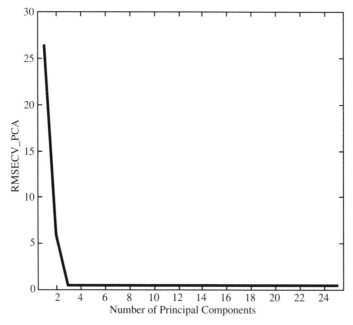

Figure 4.66. RMSECV_PCA from leave-one-out cross-validation of the complete library.

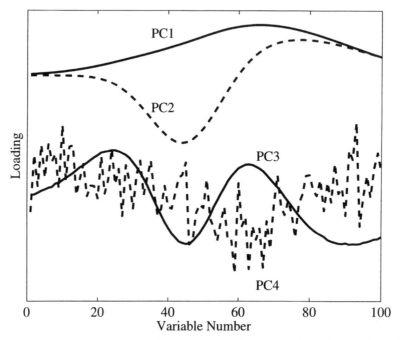

Figure 4.67. Loadings from the PCA analysis of the complete library. The offset was added for clarity.

might also be concluded that a SIMCA analysis will be problematic. However, before deriving any conclusions at this point, the third dimension must be examined.

Figure 4.68*b* displays the scores plot of score 3 vs. score 2. Class B still appears overlapped with class A, but appears more separated from C than in Figure 4.68*a*. Remember also that these score plots are only two slices of the three-dimensional object, and do not reveal all relationships. For example, classes A and C appear overlapped in Figure 4.68*b*, but not overlapped in Figure 4.68*a*. As was discussed in Habit 4 of PCA Example 1 (Section 4.2.2.1), viewing slices of the space may not reveal the full extent of overlap or separation of classes.

The conclusion drawn from this analysis is that classes A and C are separated, while class B may be overlapped with classes A and/or C. For illustrative purposes, assume that additional information is available that confirms our assertion that the unusual class B samples are actually mislabeled class A samples. These unusual class B samples will henceforth be labeled as belonging to class A. ***Caution:*** Do not make class assignments based solely on the score plots. Known class information drives the supervised methods and, therefore, it is very important that the class designations are reliable.

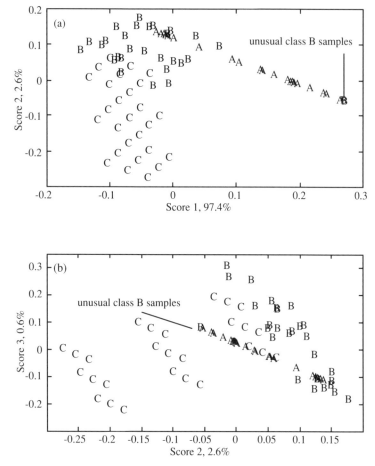

Figure 4.68. Principal component scores (*a*) 2 vs. 1 and (*b*) 3 vs. 2 for the PCA of the entire library.

After examining the training set classes with PCA, the next step is to construct and validate the SIMCA models. Ideally, the SIMCA models are constructed using the training set and validated using a completely separate test set. This separate test set is usually not available and therefore it is necessary to use a cross-validation scheme. There are many procedures for splitting the data set and a few are discussed below. Keep in mind that the training set must be carefully selected such that it is a reasonable representation of the entire class. Otherwise, the parameters of the resulting SIMCA model may not be appropriate for the entire class. One safeguard is to require that ½ to ⅔ of the original data remains in the calibration set after division.

The division of the training set is conducted one class at a time. This is to insure that there is good representation of each class in the training set. If a given class is of rank 1 or 2, it is possible to hand pick a representative set of points using a score plot. If the class is of a higher dimensionality, space filling or D-optimal experimental design approaches can be used. Still another approach to selecting samples is to choose points according to some criterion like maximum Mahalanobis distance. Finally, if the data have been collected according to some formal experimental design, it may be possible to directly select a subset. For example, if a factorial design was used to collect the entire data set, a fractional factorial subset can be considered for the training and test sets.

For the example discussed here, the calibration sets for classes A and B are selected graphically, and for class C are selected as the extremes and centers of each of the three levels in the experimental design. The selection results in 15 samples in each of the calibration sets and 12 in each of the validation sets. A score plot of all samples in class A is shown in Figure 4.69 with the calibration set samples indicated by X and the validation samples indicated by O. Similarly, score plots of classes B and C with calibration and validation samples identified are shown in Figures 4.70 and 4.71, respectively.

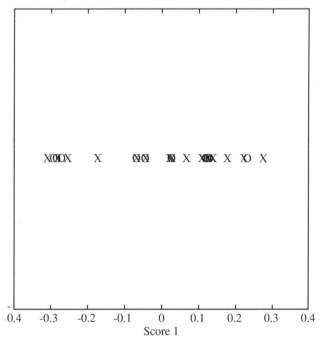

Figure 4.69. Plot of score 1 of all samples in class A (after addition of the mislabeled samples). The samples in the calibration set are Xs and the validation samples are Os.

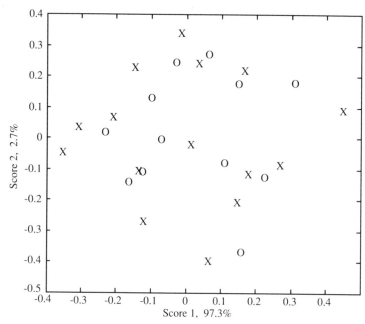

Figure 4.70. Principal component score plot of all samples in class B (after removing the mislabeled samples). The samples in the calibration set are Xs and the validation samples are Os.

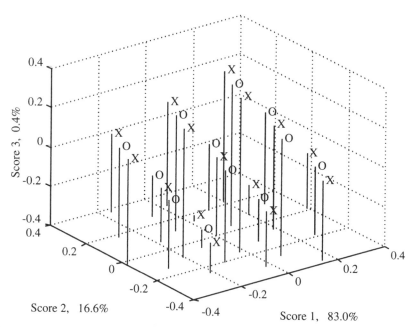

Figure 4.71. Principal component score plot of all samples in class C. The samples in the calibration set are Xs and the validation samples are Os.

149

TABLE 4.18. Summary of Classification Scenarios and Possible Root Causes

Test Set Classification	Possible Root Causes
1 Correct class—included Incorrect classes—excluded	Desired result
2 Correct class—included Incorrect classes—included	Class overlap Volume of one or multiple classes is too large
3 Correct class—excluded Incorrect class—included	Mismeasured and/or mislabeled sample (outlier)
4 Correct class—excluded Incorrect classes—excluded	Rank not correct (PCA residual large) Class volume too small (distance from box large) Statistical assumptions not obeyed Sample has another source of variation

When the PCA analysis is completed and the calibration and validation sets are chosen, the next step is to create SIMCA models for the calibration set samples. The initial rank estimates from PCA and software default class volumes are used, the performance of the models on the test samples is examined, and the SIMCA settings are adjusted as necessary.

There are many results to be reviewed because there are multiple classes for which SIMCA models are constructed and validated. The order in which to examine the results is a matter of preference, and many approaches are equally appropriate. We will review one SIMCA model at a time, and examine the test set predictions for that one model against samples from all classes. Ideal performance of a SIMCA model means that it includes as part of the class those samples that truly belong to the class and excludes those samples that are from all of the other classes. In reality, a number of classification scenarios are possible. Table 4.18 lists the possibilities along with possible root causes for misclassified test samples.

Returning to the example, the SIMCA model for the three classes are validated against the test samples from all classes beginning with the model for class A. Several diagnostic tools are discussed below and a summary is found at the end of the section in Table 4.22. These tools are used to investigate two aspects of the data set: the model and the samples. The headings for each tool indicate the aspects that are studied with that tool. The primary use of the model diagnostic tools is to determine the appropriate values for rank and boundary settings. The sample diagnostics are used to flag validation samples that are not classified correctly. There are other diagnostics available in various software packages that are not discussed in this book.

Class A SIMCA Model Validation—F_{calc} Values (Model and Sample Diagnostic): The most direct procedure for performance evaluation of the SIMCA models is to examine how well the models classified the samples from all of

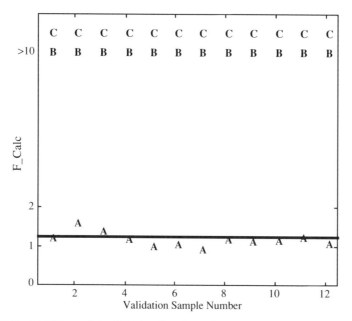

Figure 4.72. SIMCA model validation for class A. The letters indicate the class in which the validation sample is known to be a member.

the test sets. The Fcalc values for test samples from all three classes using the SIMCA model constructed for class A are plotted in Figure 4.72. The horizontal line represents Fcrit for the 95% confidence level as calculated by the software package used.

The F_{calc} values for the class A test samples are for the most part less than the critical value of 1.2, indicating that these test samples are correctly classified as belonging to class A. There are two samples that are above the critical line, even though it is known that the samples belong in this class. At a 95% confidence level, it is not unusual to have 1 out of 20 samples incorrectly classified, but here there are 2 out of 12. This observation is considered below, when the F_{calc} values are partitioned into the contribution from the PCA residual and the distance from the SIMCA boundary.

The F_{calc} values for test samples from class B range from 10.8 to well over 10,000. For the class C samples, the F_{calc} values range from 10,000 to 80,000. For all of these test samples, F_{calc} is much larger than F_{crit} and, therefore, are not considered to be members of class A. This is a desirable result that suggests that class A is not overlapped with either class B or class C.

PCA Residual and Distance from Boundary Table (Model and Sample Diagnostic): The samples that were identified in the F_{calc} plot as needing further investigation are the two class A samples that are not predicted as belonging to class A (i.e., their F_{calc} values are larger than F_{crit}). Referring to Table

4.18, this may be because the rank is not correct, the class volume is smaller than it should be, the statistical assumptions are not obeyed, or the samples have another source of variation.

To understand what might cause the misclassification, it is instructive to independently examine the two parts of F_{calc}. (Recall from Figure 4.58 that F_{calc} has contributions from the PCA residuals and the distance of the sample from the SIMCA boundary.) These are shown in Table 4.19 for the two samples in the class A test set that have F_{calc} values larger than the critical value.

Independently examining the different contributions to F_{calc} can help determine why a sample is excluded from a particular class. The PCA contribution reflects structure in the residual spectrum, which is an indication of additional sources of variation present in the unknown measurement vector (e.g., increased noise level, an unmodeled interferent, or a noise spike). The distance contribution becomes significant when the *magnitude* of the features in the unknown are unlike the training set data. This can occur when additional sources of variation are present or when the concentrations of the expected components are outside the training set range.

Determining whether the PCA or distance contributions are significantly larger than expected is dependent on the criteria used to define the SIMCA boundary. One common approach is to use conservative (large) boxes and all future samples belonging to a class are expected to fall within the box. In this case, any nonzero distance term is larger than expected. Because the expected value of the distance term is 0, the PCA term is acceptable as long as it is less than or equal to F_{crit}. (This follows from the equation $F_{calc} = PCA + distance \leq F_{crit}$ where distance = 0.)

The software package we used defaults to large boxes (see Figure 4.74) and, therefore, the distance terms found in Table 4.19 are as expected (0). The zero distance contributions indicate that the misclassification is not caused by a class volume that is too small. Referring again to Table 4.18, the other possible problems include incorrect rank, a violation of statistical assumptions, or other sources of variation. Because these data are simulated, it is known that the distribution of the residuals meet the statistical assumptions, the rank of the data is 1, and no other sources of variation are present. The other factor that has an affect on the F test is the number of degrees of freedom. There has been a fair amount of debate in the literature as to the appropriate choice of

TABLE 4.19. PCA Residual and SIMCA Boundary Distance Contributions to F_{calc} for Misclassified Class A Samples (F_{crit} = 1.2)

Validation Sample No.	PCA Contribution to F_{calc}	Distance Contribution to F_{calc}	F_{calc}
2	1.6	0.0	1.6
3	1.4	0.0	1.4

degrees of freedom. Given this ambiguity, we recommend using the critical F values as a guide rather than giving a hard and fast selection rule. For this example, it is acceptable that F_{calc} values for the class A samples are slightly larger than the critical value. A problem is indicated only when samples from the other classes have similarly close F_{calc} values.

Class B SIMCA Model Validation—F_{calc} Values (Model and Sample Diagnostic): Figure 4.73 displays the F_{calc} values for all of the test samples using the SIMCA model for class B. The F_{calc} values for class B samples 8–11 are slightly larger than F_{crit}, but are acceptably close to the critical value and are considered to be correctly classified. There are a number of the test samples known to be in class A that are predicted to be in class B (i.e., class A samples 1–5 have F_{calc} values less than the critical value of 1.22). Using the class A SIMCA model, these same samples were correctly classified as belonging to class A (see Class A SIMCA Model Validation). Referring to item 2 in Table 4.18, either there is class overlap or the volume of one or multiple classes is too large. Figure 4.68 and a three-dimensional score plot (not shown) clearly shows the partial overlap of classes A and B and there is no indication that the volume of the box is inappropriate. If a three-dimensional graphical analysis is not available, the volume of the box can be reduced and the classification of the test set repeated. An adjustment to the box size is warranted if reducing the box size results in correct classification.

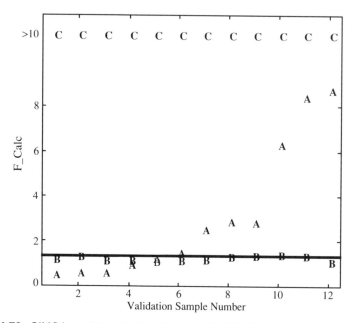

Figure 4.73. SIMCA model validation for class B. The letters indicate the class in which the validation sample is known to be a member.

If class overlap is the problem, reducing the volume size will eventually result in misclassification of samples that truly belong to the class (excluded) while samples that are not part of the class will continue to be misclassified (included). This is an indication that the root cause of the problem is nonselectivity of the measurement system. In this situation, no adjustment of the SIMCA parameters will result in unambiguous classification.

Because classes A and B were seen to be overlapped in the PCA analysis, it is not surprising that class A samples are classified as B. However, it may be confusing that class B samples are not also classified as A when examining the class A SIMCA model. (Recall the very large F_{calc} values ranging from 10.8 to 10,000.) How can the classes be overlapped and yet only have misclassification in one direction? This can be explained by examination of the PCA results from the entire training set (Figure 4.68). This shows class B as a plane and class A as a line which is partly in the plane formed by class B. In examining this figure, it is not surprising that class A samples are classified into class B; class B samples are not classified into class A because the model for class A is defined by a very narrow box. Therefore, the probability of a class B sample falling in the box spanned by class A is very small.

All of the class C validation samples are predicted to be outside of the class B SIMCA model. The minimum F_{calc} values are on the order of 100 and the maximum value is over 3000. Based on these results, it does not appear that there is any overlap of classes C and B.

PCA Residual and Distance from Boundary Table (Model and Sample Diagnostic): The different contribution to the F_{calc} value in Table 4.20 can be examined to understand why the four class B samples are not classified into class B.

The distance term for all of the misclassified B samples is zero; this indicates that the box size is not the problem. To graphically illustrate this point, a two-dimensional PCA plot for the class B samples is shown in Figure 4.74. It is clear that the class limits are adequate because the validation samples all fall within the boundary. The reason for the larger than expected F_{calc} value for these four samples is the contribution from the PCA residual. However, the

TABLE 4.20. PCA Residual and SIMCA Boundary Distance Contributions to F_{calc} for Misclassified Class B Samples ($F_{crit} = 1.2$)

Validation Sample No.	PCA Contribution to F_{calc}	Distance Contribution to F_{calc}	F_{calc}
8	1.3	0.0	1.3
9	1.3	0.0	1.3
10	1.3	0.0	1.3
11	1.3	0.0	1.3

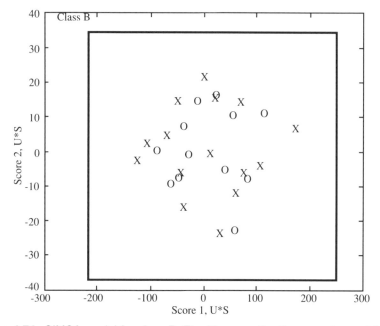

Figure 4.74. SIMCA model for class B. The Xs are calibration samples and the Os are validation samples. The solid rectangle represents the box boundary for this SIMCA model.

F_{calc} values are not extreme and, therefore, these samples are considered to be classified correctly.

Class C SIMCA Validation: F_{calc} Values (Model and Sample Diagnostic): The F_{calc} values for all of the test sets using the class C SIMCA model are plotted in Figure 4.75. The F_{calc} values for the class C test samples are found to be acceptably close to the F_{crit} value of 1.22.

The F_{calc} values for the class A test samples using the class C SIMCA model are all greater than the critical value, but a few are fairly close (minimum F_{calc} value = 2.1). Samples 1–3 for class A have low F_{calc} values, while the rest have high F_{calc} values. The conclusion is that one end of the line formed by class A is very close to class C. Because class A is a thin box, an unknown that is classified to be in both classes A and C, it is probably a class A sample. This is because the probability of having a class C sample lie within the small box around the line spanned by class A is very low.

The F_{calc} values for the class B test samples using the class C SIMCA model range from 3.9 to 7.2 indicating the class C SIMCA model can adequately distinguish between the class C and B samples.

PCA Residual and Distance from Boundary Table (Model and Sample Diagnostic): The partitioned F_{calc} values are shown for the class C validation samples with $F_{calc} > F_{crit}$ in Table 4.21.

Because all the distance contributions are zero, the boundary appears to be adequate and the reason for the larger than expected F_{calc} value is due to the PCA residual. Again, the conclusion is that these class C samples have acceptable F_{calc} values.

Summary of Validation Diagnostic Tools for SIMCA

The results from the application of the validation diagnostic tools for this example are as follows:

1. Classes A, B, and C have ranks of 1, 2, and 3, respectively.
2. Classes A and B are overlapped.
3. Part of class A is close to class C.
4. The statistical assumptions are not strictly obeyed (it appears that the degrees of freedom are not accurate).

If the overlap of classes A and B is not acceptable, additional discriminating measurements are required. If samples have an F_{calc} value larger than the critical value and it is believed to be related to assumption violations, the critical value should be increased. Our advice is to choose a larger critical value such that all validation samples are predicted in the correct classes. One way to empirically determine a critical value is to take the average of the F_{calc} values from the correctly classified samples and add some multiple of the standard deviation of the F_{calc} values. Usually a multiple of 3–5 is adequate. With these data, a multiple of 3 will give an F_{calc} of 1.6 for all three classes which avoids the erroneous exclusion of test samples from their true classes. Choosing this larger critical value does not take care of the problem of overlap (i.e., erroneous inclusion of test samples into the wrong classes).

Table 4.5 in Section 4.2.2.1 and Table 4.22 below summarize the validation diagnostic tools discussed in this section. The first column in the table lists the name of each tool and the second column describes results from both well-behaved and problematic data.

TABLE 4.21. PCA Residual and SIMCA Boundary Distance Contributions to F_{calc} for Misclassified Class C Samples ($F_{crit} = 1.22$)

Sample No.	PCA Contribution to F_{calc}	Distance Contribution to F_{calc}	F_{calc}
4	1.5	0.0	1.5
6	1.3	0.0	1.3
7	1.3	0.0	1.3

TABLE 4.22. Summary of Validation Diagnostics for SIMCA

Diagnostics	Description and Use
Model	
F_{calc} values (F_{calc} vs. sample number)	F_{calc} is a quantitative measure of the reliability of a classification of a sample into a class. An F_{calc} smaller than the critical value (F_{crit}) indicates that the sample belongs to that class.
	Used to assess the quality of a SIMCA model.
	For ideal data, the F_{calc} values are consistent with known class membership. All samples that are known to belong to a given class are classified in that class while other samples are excluded.
PCA residual and distance from boundary table	The F_{calc} value for the classification of a sample is partitioned into a portion due to the PCA residuals and another part due to the distance from the class boundary.
	Used to help determine optimum rank and class boundary for the SIMCA model.
	The PCA residuals should be small for samples that belong to the class and large otherwise. The distance from the boundary should be zero for all validation samples that belong to the class.
Sample	
F_{calc} values (F_{calc} vs. sample number)	F_{calc} is a quantitative measure of the quality of a classification of a sample into a class. An F_{calc} smaller than the value (F_{crit}) indicates that the sample belongs to that class.
	Used to identify outlying samples as those with unusually large F_{calc} values.
PCA residual and distance from boundary table	The F_{calc} value for the classification of a sample is partitioned into a portion due to the PCA residuals and another part due to the distance from the class boundary.
	Used to investigate why an outlying sample has a large F_{calc} value.
Variable	No variable diagnostics are discussed.

Three "What If" scenarios are now presented using the data discussed in Example 1. These include: (1) changing class volume, (2) choosing an incorrect rank, and (3) mislabeling samples in the training set.

"What If" 1. Changing Class Volume. Consider using the class C samples to construct a SIMCA model and using the class A validation samples as a test set. Remember that class C is three-dimensional and A is one-dimensional and one end of the line lies close to class C. The prediction results using the default class size have been discussed and the F_{calc} values are plotted in Figure

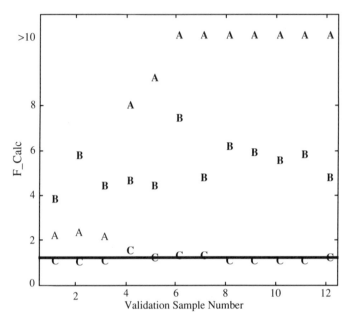

Figure 4.75. SIMCA model validation for Class C. The letters indicate the class in which the validation sample is a member.

4.75. All of the class A samples are correctly excluded from class C. For this scenario, the size of the box for the class C SIMCA model is increased and the resulting F_{calc} values for the class A test samples are examined.

Table 4.23 lists the F_{calc} values for the 12 class A test samples using three different class C SIMCA models. Column one lists the class A sample number and column two displays the PCA contribution to F_{calc}. This value does not change as the box size is changed. Column 3 displays the distance contribution to F_{calc} when the default box size is used and column 4 displays the F_{calc} (PCA + distance contributions). Columns 5-6 and 7-8 list the same values for large and larger box size models, respectively. The first observation from this table is that as the class volume increases, the F_{calc} values for all of the class A test samples decrease. This means these samples are closer to being classified as members of class C. However, for these three different class C models, all of the F_{calc} values for the class A test samples remain larger than the critical value of 1.6 and therefore are not classified as belonging to class C. In fact, because the PCA contribution to F_{calc} for all of the samples is above the critical value, none of these samples will be classified as class C samples no matter how large the box is made. It must therefore be the case that the samples from class C and class A exist in different spaces.

For this example, the size of the box is manipulated to illustrate a point. In general, the size of the box is set such that all samples are correctly included and excluded during the validation phase. If the SIMCA model using a default box size includes samples from other classes, smaller boxes can be úsed in an attempt to exclude these samples.

TABLE 4.23. Effect of Increasing the Box Size of Class C SMICA Model on Class A Samples ($F_{crit} = 1.2$)

No.	PCA Contribution to F_{calc}	Default Distance Contribution to F_{calc}	Total F_{calc}	Large Distance Contribution to F_{calc}	Total F_{calc}	Larger Distance Contribution to F_{calc}	F_{calc}
1	2.0	0.1	2.1	0.0	2.0	0.0	2.0
2	2.1	0.2	2.3	0.0	2.1	0.0	2.1
3	1.9	0.2	2.1	0.0	1.9	0.0	1.9
4	3.7	4.3	8.0	0.3	4.0	0.0	3.7
5	3.5	5.6	9.1	0.7	4.2	0.0	3.5
6	5.0	11.6	16.6	3.6	8.6	0.0	5.0
7	6.5	23.1	29.6	11.1	17.6	0.0	6.5
8	6.7	23.9	30.6	11.6	18.3	0.0	6.7
9	7.0	24.7	31.7	12.1	19.1	0.0	7.0
10	10.0	44.9	54.9	25.9	35.9	0.0	10.0
11	12.6	58.8	71.4	36.1	48.7	0.0	12.6
12	11.1	59.5	70.6	36.7	47.8	0.0	11.1

"What If" 2. Choosing An Incorrect Rank. The performance of the class C SIMCA model, previously determined to be a rank three model, is examined using a rank of two. A rank three SIMCA model appropriately includes class C samples and excludes class A and B samples. When a rank two SIMCA model is estimated, two class C test samples are incorrectly excluded. An examination of F_{calc} reveals that the distance term is zero and the PCA residual is the only contribution. This indicates a modeling problem. When this SIMCA model is applied to the class A and B test samples, three and eight samples are incorrectly included as part of class C, respectively.

The reason for these misclassifications is that the training sample residuals for the rank two SIMCA model are larger than the rank three model. Therefore, prediction samples can have a larger residual and still be considered to be members of class C.

If too large a rank is selected, there is a chance of overfitting. This is not desirable because it will impact long-term performance of the model. This is demonstrated in SIMCA Example 2 below.

"What If" 3. Mislabeling Samples in the Training Set. For the third "what if," the class B SIMCA model is considered, only this time it is constructed from a training set that includes the two mislabeled class A samples. When this new model is applied to the classification of the class A samples, all are classified as members of class B (i.e., there is total overlap). When the mislabeling was corrected, only five out of 12 of the class A test samples were classified into class B (see Figure 4.73). The mislabeled samples have significantly decreased the ability of the model to distinguish between classes A and B. Using this SIMCA model, the (incorrect) conclusion would be that the measurements are not selective enough to distinguish between classes A and B.

Habit 5. Use the Model for Prediction

The final step of constructing the SIMCA models is to merge the calibration and test samples for each of the classes and reconstruct new SIMCA models using all of the data. The rank and boundary parameters determined in Habit 4 are used for the final models. These models are used to predict the class(es) of unknown samples. Table 4.24 contains the F_{calc} values for three unknown samples where the empirically determined critical value is 1.6. From the F_{calc} values, the conclusions are that unknown 1 is not a member of any class in the training set, unknown 2 is a member of class B, and unknown 3 is a member of both classes A and B.

Habit 6. Validate the Prediction

Several prediction validation tools are discussed below and a summary is found at the end of the section in Table 4.25.

F_{calc} *Values:* The process of classifying a sample by comparing F_{calc} with F_{crit} has already validated the predictions to some extent. A very large F_{calc} value (e.g., unknown 2 on class A has a F_{calc} of 11,000) indicates a very large difference between the unknown sample and the calibration samples from that class. Unknown 1 has large F_{calc} values for all of the classes with most of the contribution coming from the PCA measurement residual. Unknown 2 is within the box of all SIMCA models but has large PCA contributions with classes A and C. Unknown 3 is excluded only from class C primarily because of a large contribution from the distance term (the expected value is zero).

Measurement Residual Plot: There are residual plots for each unknown sample for every SIMCA model. The residual spectra for samples that belong to a class are expected to resemble in magnitude and shape normally distributed noise as found in the training set. Depending on the structure of the residuals, it may be possible to identify failures in the instrument (e.g., excessive noise) or chemical differences between the calibration and unknown samples (e.g., peaks in the residuals). The residual plot may help identify why a sample is not classified into any given class.

TABLE 4.24. PCA and Distance Contributions to F_{calc} for Three Unknowns Using Class A, B, and C Models ($F_{crit} = 1.6$)[a]

| Unk No. | Class A | | | Class B | | | Class C | | |
	PCA F_{calc}	Distance F_{calc}	F_{calc}	PCA F_{calc}	Distance F_{calc}	F_{calc}	PCA F_{calc}	Distance F_{calc}	F_{calc}
1	84	0.0	84	15	0.0	15	41.1	3.1	44.2
2	11,000	0.0	11,000	1.1	0.0	**1.1**	4.1	0.0	4.1
3	1.1	0.0	**1.1**	0.6	0.0	**0.6**	1.3	1.3	2.6

[a]F_{calc} values less than F_{crit} are in boldface type.

The residual spectra for unknown 1 for classes A, B, and C are shown in Figure 4.76. The spectra have been offset for clarity and the three pairs of horizontal lines indicate the range of the PCA residuals for the training set data for each of the models. Consistent with the large F_{calc} values for unknown 1 shown in Table 4.24, the residual spectra have features that are considerably larger than those found in the training set. The nonrandom behavior of the residual spectra suggests that another chemical component is present in this unknown sample. The size of the residual relative to the calibration range also corresponds to the relative sizes of the F_{calc} values. If any of these results are unexpected, it is advisable to double check the prediction by repeating the measurement and the SIMCA prediction.

For unknown 2, the contribution of the residual to F_{calc} for class A is very large, small for class B, and roughly double the critical value for class C. The residuals shown in Figure 4.77 are consistent with this. There appears to be a component in this unknown that is not present in class A, as is indicated by the very large structured residual. The class B residual is within the range of the calibration spectra residuals, and is slightly outside the range of calibration sample residuals for class C. There is no structure in the residuals for class C and the noise level is only slightly increased. It is therefore difficult to make a conclusive statement as to why unknown 2 is not a member of class C.

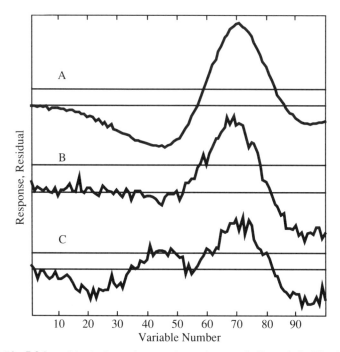

Figure 4.76. PCA residuals for unknown 1 on classes A, B and, C. The horizontal lines show the range of the calibration residuals for the respective classes.

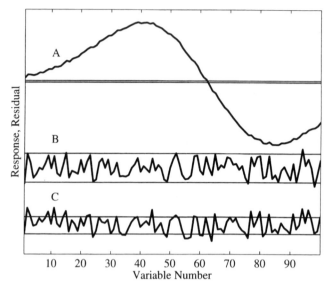

Figure 4.77. PCA residuals for unknown 2 on classes A, B and, C. The horizontal lines show the range of the calibration residuals for the respective classes.

Unknown 3 was classified as a member of classes A and B. The residual spectra for classes A and B (see Figure 4.78) are largely within the range of the calibration sample residuals. The class C residuals are slightly larger in magnitude than the calibration residuals, but there are no obvious systematic features. With a contribution of 1.3, the residuals alone are not enough to keep the sample from being classified into class C and the distance contribution indicates that the sample is outside of the range of the training set. This can be verified by comparing the raw plot of this unknown with the training set samples for class C (compare Figure 4.79 below with Figure 4.61).

Raw Measurement Plot: A plot of the three unknown spectra is shown in Figure 4.79. All of the unknown spectra look reasonable when compared to the training set spectra (Figure 4.61). There are no obvious features that would result in larger than expected F_{calc} values. This is not surprising because subtle differences in shape can be difficult to see in these plots. These small differences can, however, have a significant effect on the PCA residuals and/or the distance of the samples from the SIMCA box.

Scores Plot: The score plot displays the relationship of the samples to each other in row space. It does not show the residual information and only contains that fraction of the total variation that is described by the PCs that are examined. There are a series of score plots (score 2 vs. score 1, score 3 vs. score 2, etc.) available for each unknown and each class model.

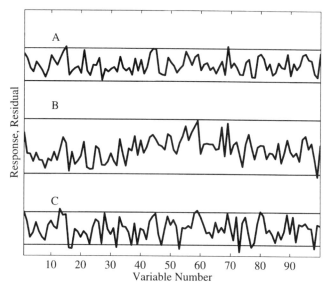

Figure 4.78. PCA residuals for unknown 3 on classes A, B and, C. The horizontal lines show the range of the calibration residuals for the respective classes.

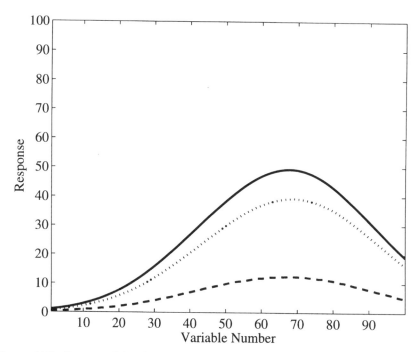

Figure 4.79. Spectra of the three unknown samples. Solid, unknown 1; dashed, unknown 2; dotted, unknown 3.

Shown in Figure 4.80 are the scores of class A, class B, and unknown 3 plotted in the score 2 vs. score 1 plot of class B. Unknown 3 is located in the same general area as the samples from class A in the overlap region of classes A and B. From the small PCA contribution to F_{calc} it is known that unknown 3 does not lie far out of this plane. This plot confirms what the other diagnostics tools have indicated regarding this unknown sample.

Summary of Prediction Diagnostic Tools for SIMCA: From the prediction diagnostics it was found that unknown 1 is not a member of any of the classes, unknown 2 is a member of class B, and unknown 3 is a member of classes A and B.

Table 4.25 summarizes the prediction validation tools discussed in this section. The first column in the table lists the name of each tool and the second column describes results from both well-behaved and problematic data.

4.3.2.2 SIMCA Example 2   SIMCA is applied to the same data set that was presented in PCA Example 2 (Section 4.2.2.2). An array of Taguchi-type semiconductor sensors doped with various catalysts were considered for gas-phase monitoring. In the investigation, 10 organic species with concentrations varying from 0–3 ppt were studied with seven sensors. The intensity and maximum slope were measured for each sensor resulting in a total of 14 measurements for each sample analyzed.

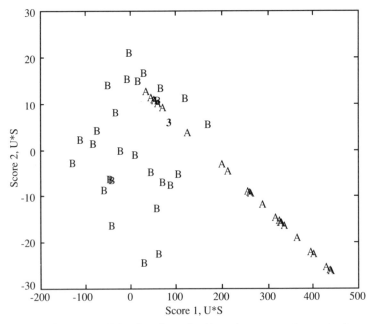

Figure 4.80. PC2 versus PC1 for class B. Also shown are the projections of the class A calibration samples and unknown 3 onto this plane.

TABLE 4.25. Summary of Prediction Diagnostics for SIMCA

Prediction Diagnostics	Description and Use
F_{calc} values	F_{calc} is a quantitative measure of the quality of a classification of a sample into a class. An F_{calc} smaller than the critical value (F_{crit}) indicates that the sample belongs to that class.
	Used to determine and help assess the reliability of the classification of an unknown sample.
	The F_{calc} value for the classification of a sample is partitioned into a portion due to the PCA measurement residuals and another part due to the distance from the class boundary.
	The PCA measurement residuals should be small for samples that belong to the class and large otherwise. The distance from the boundary should be zero for samples that belong to the class.
Measurement residual plot $[(\mathbf{r} - \hat{\mathbf{r}})$ vs. variable number]	The residuals are that portion of the sample measurement vector not fit by the PCA model for a given class.
	Used to help assess the reliability of the classification of unknown samples.
	The classification is suspect when the residual vector is significantly different from the training sample residuals.
	Examining the variables where a sample has unusual features can be used to determine the cause of the problem.
Raw measurement plot (**r** vs. variable number)	A plot of the measurement vector for the prediction sample.
	Used in conjunction with knowledge of the chemistry and the measurement system to assign cause when suspect classifications are encountered.
Scores Plot [PCy vs. PCx (vs. PCz)]	The scores shows the location of the prediction sample relative to the training set samples. The plots are 2 or 3 dimensional representations of the row space.
	Used to investigate samples that have been identified as being unusual.

The conclusion of the unsupervised PCA analysis was that there was enough selectivity to distinguish between compounds based on functional groups. However, it was unclear whether the sensor array could distinguish between compounds with the same chemical functionality. SIMCA models for 2 of the 10 compounds, triethylamine (TEA) and methylethylketone (MEK), are constructed and validated against the entire data set containing all 10 classes of compounds.

Habit 1. Examine the Data
All of the data are plotted in Figure 4.34 of Example 2 in Section 4.2.2.2. Nothing unusual is seen in Figures 4.81a and b which display the raw data (sensor intensities and slopes) for TEA and MEK, respectively.

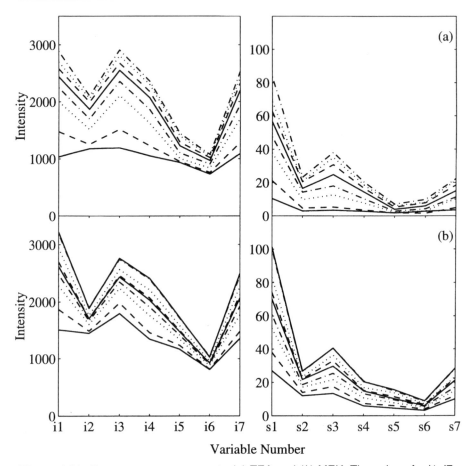

Figure 4.81. Sensor array responses to (*a*) TEA and (*b*) MEK. The values for i1–i7 are the sensor intensities and s1–s7 are the sensor slopes.

Habit 2. Preprocess as Needed

Autoscaling is applied because the intensities and slopes are of different magnitudes, as seen on the y axis in Figure 4.81.

Habit 3. Estimate the Model

The procedure from Table 4.16 was followed to construct the SIMCA models.

Habit 4. Examine the Results / Validate the Model

The PCA results for the two individual classes are examined first. The PCA of the entire training set is performed in PCA Example 2 in Section 4.2.2.2. In the current PCA analysis, the ranks and boundaries are chosen and then the SIMCA models are constructed and validated.

PCA of TEA—Percent Variance Plot (Model Diagnostic): For the TEA class the first through fourth PCs describe 97.3%, 2.2%, 0.2%, and 0.1% of the variation, respectively. This suggests that a rank of two is appropriate, assuming the noise in the data is more than 0.2% of the variance.

Measurement Residual Plot (Model, Sample, and Variable Diagnostic): The residuals after the first PC, the first two PCs, and the first three PCs are shown in Figures 4.82a–c, respectively. These residuals have been converted back to the original scale ("unautoscaled") to facilitate comparisons to the original data. The residuals after three PCs do not appear to be much smaller than the residuals after two, indicating that a two-PC model may be appropriate.

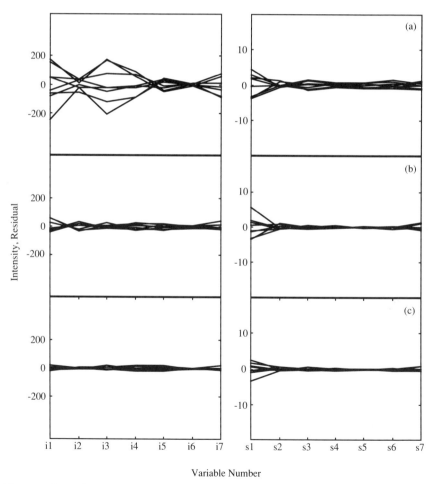

Figure 4.82. PCA residuals for TEA after (*a*) 1 PC, (*b*) 2 PCs, and (*c*) 3 PCs. The values for i1–i7 are the sensor intensities and s1–s7 are the sensor slopes.

Root Mean Square Error of Cross Validation for PCA Plot (Model Diagnostic): Figure 4.83 shows that the RMSECV_PCA decreases significantly after the first and second PCs are added, but the decrease is much smaller when additional PCs are added. This implies that a two-component PCA model is appropriate.

Loadings Plot (Model and Variable Diagnostic): The first three loading vectors from the PCA models for the TEA class are shown in Figure 4.84. Because the measurements are not continuous, it is not possible to use the randomness of the loadings to help select the rank. The loadings on the first PC are all approximately 0.26. The first principal component describes mostly concentration changes (see the scores plot below and the discussion in Section 4.2.2.2). This nonvarying behavior of loading 1 is due to the fact that the data have been autoscaled, and therefore the sensitivity to concentration changes are nearly equivalent for all variables.

Scores Plot (Sample Diagnostic): The score plots are shown in Figure 4.85 where the samples are labeled with the corresponding concentrations. The first two PCs plotted with the concentration information indicates some nonlinearity in the signal with concentration. The data form a curved, rather than straight, line in two-dimensional space, and it is known that there is only one chemical component in the samples. From Figure 4.85*b,* it does not appear as if the third principal component describes any simple behavior for this collection of samples. That is, no simple trend with concentration is visible, nor is there any clustering of the samples. With this few samples, it is important to avoid overfitting. The conclusion is that PC3 does not describe systematic variation, and will therefore not be included in the SIMCA model.

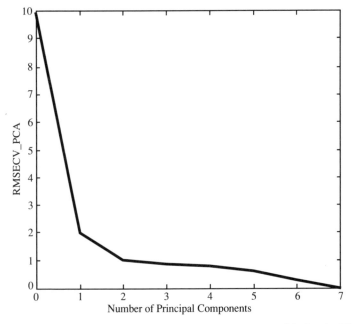

Figure 4.83. RMSECV_PCA from leave-one-out cross-validation for TEA.

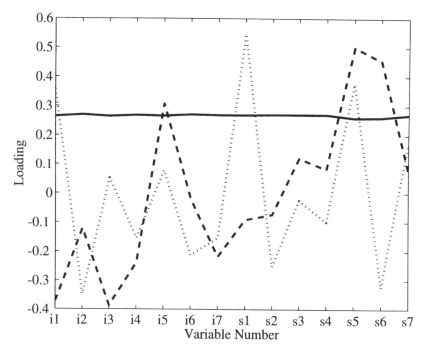

Figure 4.84. Loadings plot for the PCA model of TEA. Solid, PC1; dashed, PC2; dotted, PC3.

PCA of MEK—Percent Variance (Model Diagnostic): The first through fourth PCs of the MEK data describe 99.0%, 0.4%, 0.3%, and 0.1% of the variation, respectively. Assuming that the noise is greater than 0.4% of the variation, a one-component PCA model may be appropriate.

Measurement Residual Plot (Model, Sample, and Variable Diagnostic): The residuals after two PCs (not shown) are not much smaller than the residuals after one PC, indicating that a one-PC model may be appropriate.

Root Mean Square Error of Cross Validation for PCA Plot (Model Diagnostic): Figure 4.86 shows that the RMSECV_PCA decreases significantly after the first PC is added, but the decrease is much smaller when additional PCs are added. This implies that a one-component PCA model is appropriate.

Loadings Plot (Model and Variable Diagnostic): The loadings of the first two PCs for the MEK show similar behavior to the loadings from the TEA data and are not helpful in determining rank.

Scores Plot (Sample Diagnostic): The score plot of score 2 vs. score 1 is shown in Figure 4.87, with the samples labeled with the corresponding concentrations. The scores on the first PC are related to concentration (the score value increases as the concentration increases) but there is no strong dependence with respect to score 2. This supports a one-component PCA model as being appropriate.

PCA on Entire Training Set: See PCA Example 2, Section 4.2.2.2.

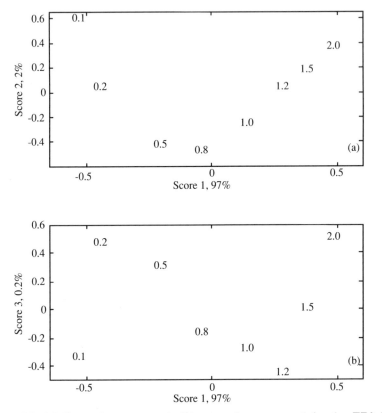

Figure 4.85. (*a*) Score 2 vs. score 1, (*b*) score 3 vs. score 1 for the TEA PCA model. The samples are labeled with concentration (ppt).

From the PCA analysis, it was concluded that appropriate ranks for the TEA and MEK SIMCA models are two and one, respectively. The next step is to construct SIMCA models and test their performance on validation samples. The ranks determined during the PCA analyses and the default settings for the class volume size for the models are used.

To test the models, the training set is divided into calibration and validation sets, as shown in Table 4.26. The predictive ability of the TEA and MEK SIMCA models is then evaluated using samples from all 10 classes.

TEA SIMCA Validation—F_{calc} Values (Model and Sample Diagnostic): A rank two SIMCA model is used to generate the F_{calc} values for the three validation samples known to belong to the TEA class (see Table 4.27). As is the desired result, each of the samples have an F_{calc} value smaller than the critical value and are therefore classified as TEA samples.

The F_{calc} values can also be used to validate rank by examining the impact of different ranks on the classification of the validation samples. When the rank is too large, overfitting occurs and the validation samples will be incorrectly excluded from the class for which the SIMCA model is constructed.

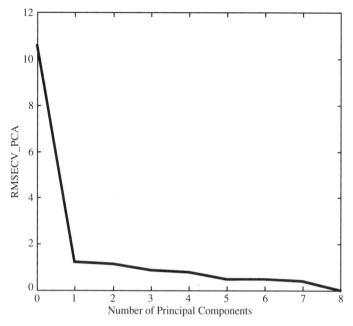

Figure 4.86. RMSECV_PCA from leave-one-out cross-validation for MEK.

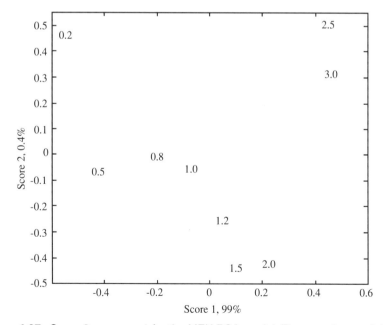

Figure 4.87. Score 2 vs. score 1 for the MEK PCA model. The samples are labeled with concentration (ppt).

TABLE 4.26. Calibration and Validation Sets for TEA and MEK

Class	Concentration of Calibration Samples (ppt)	Concentration of Validation Samples (ppt)
TEA	0.10, 0.25, .075. 1.25, 2.00	0.50, 1.00, 1.50
MEK	0.25, 0.50, 1.25, 1.50, 2.50, 3.00	0.75, 1.00, 2.00

Table 4.28 displays F_{crit} and F_{calc} for the three TEA test samples as a function of the rank of the TEA SIMCA model. For a rank of one or two, the validation samples are predicted to be in the class ($F_{calc} < F_{crit}$). When the rank is three, the third validation sample is excluded from the class. (Note that F_{crit} is a function of rank and therefore is different for each SIMCA model.) This experiment confirms the conclusion that two is a reasonable rank for the TEA SIMCA model.

Because the number of samples in the calibration set is small, all of the TEA samples (calibration and validation) are used to construct a TEA SIMCA model that is tested against the validation samples from the rest of the training set (MEK plus the other eight classes of compounds). F_{calc} values for these samples are shown in Figure 4.88a. In Figure 4.88b the data are plotted with a smaller y axis scale in order to examine values close to the critical value. The x axis corresponds to the numerical class identification shown in Table 4.6 (Section 4.2.2.2, PCA Example 2) and the samples are labeled on the plot with concentration values (ppt). (Note: TEA is class 10.) The TEA SIMCA model has correctly rejected all of the test set samples from the other classes. For samples within a class, the F_{calc} values increase as the concentration increases. That is, at a lower concentration, the test sample appears more similar to TEA than at a higher concentration. As the concentration approaches zero, it is increasingly difficult to distinguish between the different classes. This is demonstrated with the air class (0) where many of these samples have F_{calc} values close to the critical value. Overall these results indicate that the measurements are selective enough to distinguish between TEA and the other chemicals within the concentration ranges considered.

TABLE 4.27. PCA and Distance Contributions to F_{calc} for TEA SIMCA Model and TEA Validation Samples ($F_{crit} = 2.1$)

Sample No.	PCA Contribution to F_{calc}	Distance Contribution to F_{calc}	F_{calc}
1	1.0	0.0	1.0
2	0.6	0.0	0.6
3	1.5	0.0	1.5

TABLE 4.28. F_{calc} **Values for TEA Validation Samples Using SIMCA Models with Ranks One through Three**

TEA	Rank = 1	Rank = 2	Rank = 3
F_{crit}	1.9	2.1	4.0
F_{calc} test sample 1	1.3	1.0	3.2
F_{calc} test sample 2	0.6	0.6	2.0
F_{calc} test sample 3	0.4	1.5	5.5

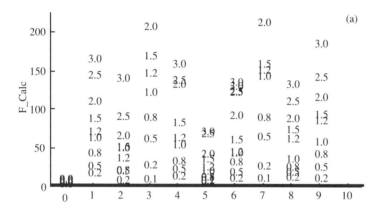

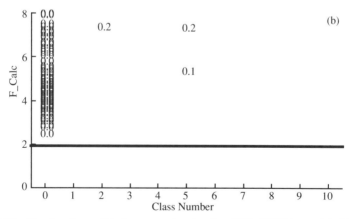

Figure 4.88. F_{calc} for the validation samples for the TEA SIMCA model. The horizontal line represents the critical F value for the SIMCA model made with the calibration data (1.92). The sample identification numbers correspond to concentration (ppt). TEA is class 10.

TABLE 4.29. PCA and Distance Contributions to F_{calc} for MEK SIMCA Model and MEK Validation Samples ($F_{crit} = 1.84$)

Sample No.	PCA Contribution to F_{calc}	Distance Contribution to F_{calc}	F_{calc}
1	1.4	0.0	1.4
2	1.5	0.0	1.5
3	1.7	0.0	1.7

PCA Residual and Distance from Boundary Table (Model and Sample Diagnostic): Because there are no samples misclassified the F_{calc} values are not examined more closely. The distance terms for the validation samples for the model (Table 4.27) are all zero. This is an indication that the class volume size is adequate.

MEK SIMCA Validation—F_{calc} Values (Model and Sample Diagnostic): Table 4.29 displays the F_{calc} values for the three validation samples known to be MEK using the MEK SIMCA model. All values are smaller than the critical value and are correctly classified as belonging to the MEK class. This is again the desired result.

Examining F_{calc} while changing rank confirms a rank one SIMCA model for MEK. Using rank one, the three validation samples are all predicted to be in the class. Using rank two or three, all validation samples are predicted to be outside of the MEK class (see Table 4.30).

All of the MEK samples (calibration and validation) are used to construct an MEK SIMCA model that is tested against the validation samples from the rest of the training set (TEA plus the other eight classes of compounds). The F_{calc} values resulting from applying the MEK SIMCA model are shown in Figure 4.89*a* and an expanded *y* axis in Figure 4.89*b*. The *x* axis is labeled with a numerical class identification summarized in Table 4.6 (Section 4.2.2.2, PCA Example 2) and the samples are labeled with concentration values (ppt). (Note: MEK is class 4.) There are two interesting observations from these results. First, the air samples (class 0) have much higher F_{calc} values with the MEK SIMCA model than with the TEA model. This is because the lowest concentration in the MEK training set samples is 0.2 ppt, while the lowest concentration in the TEA

TABLE 4.30. F_{calc} Values for MEK Validation Samples Using SIMCA Models with Ranks One through Three

MEK	Rank = 1	Rank = 2	Rank = 3
F_{crit}	1.8	2.0	2.2
F_{calc} test sample 1	1.4	3.1	6.8
F_{calc} test sample 2	1.5	3.1	7.0
F_{calc} test sample 3	1.7	2.7	5.4

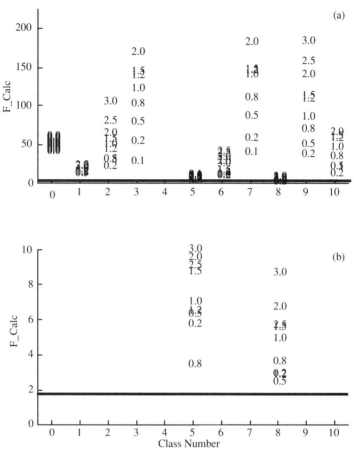

Figure 4.89. F_{calc} for the validation samples for the MEK SIMCA model. The horizontal line represents the critical F value for the SIMCA model made with the calibration data (1.75). The sample identification numbers correspond to concentration (ppt). MEK is class 4.

training set samples is 0.1 ppt. Second, classes 5 (ethyl acetate) and 8 (1-propanol) do not have as large F_{calc} values as the rest of the classes. This means that the response of the sensors to these compounds is most similar to MEK. However, all values are above F_{crit}, indicating that the sensors have sufficient selectivity to distinguish between MEK and the other organic species.

PCA Residual and Distance from Boundary Table (Model and Sample Diagnostic): Because no samples are misclassified, the breakdown of F_{calc} into the residual and distance contributions is not examined. The results from the validation samples for the model (Table 4.29) indicate that the class volume size is adequate.

Summary of Validation Diagnostic Tools for SIMCA: From the validation analysis it was found that the measurements and SIMCA models for TEA and MEK are adequate to distinguish these materials from the other chemicals considered in this study. The diagnostic tools also indicate that it is more difficult to discriminate between the classes at lower concentrations.

Habit 5. Use the Model for Prediction

The results of predictions using the two SIMCA models on four unknown samples are shown in Table 4.31. These preliminary results indicate that unknowns 1 and 4 are not a member of either class, unknown 2 is MEK and unknown 3 is TEA.

Habit 6. Validate the Prediction

F_{calc} *Values:* The samples that are excluded from the TEA and MEK classes have F_{calc} values ranging from 18 to 61. This clearly indicates that these samples do not belong to either class (critical values of 1.9 and 1.7). The exclusion of these samples is based on the PCA residual because the distance contribution is zero.

Measurement Residual Plot: The PCA residuals for the four unknowns using the TEA model are shown in Figure 4.90 with the range of the calibration residuals shown as solid horizontal lines. Unknowns 1, 2, and 4 have large residuals, which is reflected in the F_{calc} values. Unknown 3 has a small residual which is consistent with the classification as a member of the TEA class. The PCA residuals for the MEK model are shown in Figure 4.91 with the range of the calibration residuals shown as solid horizontal lines. Again, all unknowns have residuals which are consistent with the F_{calc} values.

Raw Measurement Plot: The raw data plots of the unknowns are shown in Figure 4.92. None of the samples appear to be unusual when this plot is compared to the training set raw data plots (Figure 4.81).

Scores Plot: The score plot is used to examine the location of the samples in the PCA space. A three-dimensional PCA scores plot for TEA model is shown in Figure 4.93. (Keep in mind that only the first two PCs were used to construct the TEA SIMCA model.) The TEA training set samples and the four unknowns are shown on this plot. Unknown 3, which was predicted to be

TABLE 4.31. F_{calc} for Four Unknowns for the TEA and MEK SIMCA Models[a]

Unk No.	TEA (F_{crit} = 1.9)			MEK (F_{crit} = 1.7)		
	PCA F_{calc}	Distance F_{calc}	F_{calc}	PCA F_{calc}	Distance F_{calc}	F_{calc}
1	46	0.0	46	56	0.0	56
2	21	0.0	21	1.4	0.0	**1.4**
3	0.3	0.0	**0.3**	38	0.0	38
4	60.9	0.0	60.9	18	0.0	18

[a]F_{calc} values less than F_{crit} are shown in boldface type.

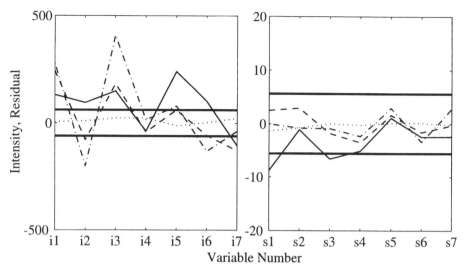

Figure 4.90. PCA residuals for unknowns from the TEA SIMCA model. Solid, unknown 1; dashed, unknown 2; dotted, unknown 3; dashed–dotted, unknown 4.

TEA, is close to the TEA training set samples. Unknowns 1, 2, and 4 are located above the plane of the TEA space. It may seem inconsistent that unknown 2 lies further above the plane than unknown 1, because the F_{calc} for unknown 2 is less than for unknown 1. However, this score plot only describes three of the possible fourteen total dimensions in row space. Unknown 1 has more residual variation in the other 11 dimensions not shown on this plot.

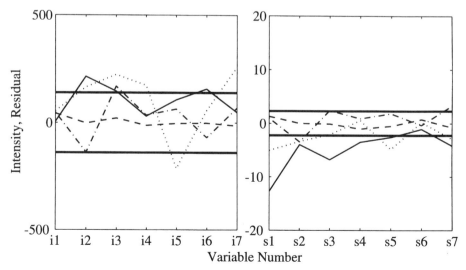

Figure 4.91. PCA residuals for unknowns from the MEK SIMCA model. Solid, unknown 1; dashed, unknown 2; dotted, unknown 3; dashed–dotted, unknown 4.

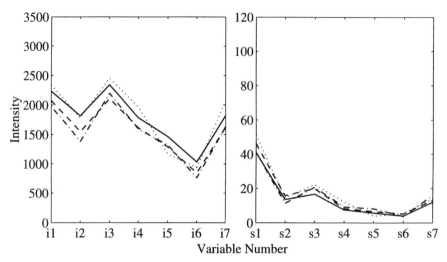

Figure 4.92. Plot of prediction samples. Solid, unknown 1; dashed, unknown 2; dotted, unknown 3; dashed–dotted, unknown 4. The values for i1–i7 are the sensor intensities and s1–s7 are the sensor slopes.

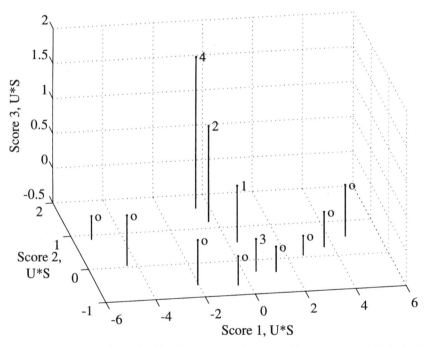

Figure 4.93. PCA of TEA SIMCA library samples (○) with unknowns (labeled with numbers). The first two principal components are used to make the TEA SIMCA model.

The score plot for the MEK training set samples and unknowns 1–4 using the MEK SIMCA model is shown in Figure 4.94. The MEK model is one dimensional and the first PC is shown on the plot. Unknown 2, which was classified as MEK, is located close to the MEK training set samples. Unknowns 1 and 4 are within the SIMCA box (defined by the ends of the line) but are clearly off the line (large PCA contribution to F_{calc}). Unknown 3 appears to be quite close to the MEK training set data but has a large F_{calc} value. This indicates that unknown 3 has significant residual variation in one or more of the other 12 dimensions not shown on this plot.

Summary of Prediction Diagnostic Tools for SIMCA: From the prediction diagnostics, the conclusion is that unknowns 1 and 4 do not belong to either of the TEA or MEK classes. Sample 3 is a member of the TEA class and sample 2 is a member of the MEK class. There is considerable reliability in the classifications due to the large F_{calc} values for the excluded samples both in the validation and prediction phases. The residuals and score plots are consistent with the F_{calc} values.

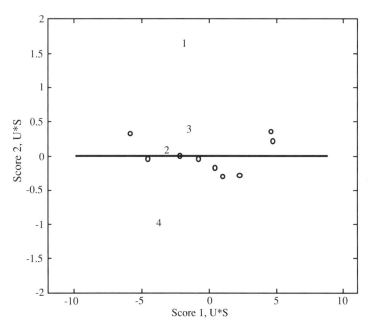

Figure 4.94. PCA of MEK SIMCA library samples (○) with unknowns (labeled with numbers). The one-component MEK SIMCA model (with boundaries) is shown by the solid line.

4.3.3   Summary of Supervised Pattern Recognition Methods

Supervised pattern recognition methods are used for predicting the class of unknown samples given a training set of samples with known class membership. Two methods are discussed in Section 4.3, KNN and SIMCA.

Summary of KNN

K-nearest neighbor classification is a general approach for classifying unknown samples. The predicted class of an unknown is assigned to the class of the sample(s) lying nearest to it in multidimensional space.

Weaknesses

1. The classical KNN approach does not have outlier detection capabilities. That is, a classification is always made, whether or not the unknown is a member of any of the classes in the training set. In Section 4.3.1, the method presented includes outlier diagnostics which are generally not present in commercial statistical software.

2. KNN does not take advantage of class shape information.

Strengths

1. KNN is a simple technique to implement.

2. It is an appropriate method when there are few samples per class (in contrast to SIMCA).

3. The training set can easily be updated with additional samples. (If a new class is added, be sure to reevaluate the choice of K and the goodness criterion critical value.)

Summary of SIMCA

While KNN only uses physical closeness of samples to construct models, SIMCA uses the position and shape of the object formed by the samples in row space for class definition. Modeling the object formed by an individual class is accomplished with principal components analysis (PCA) (see Section 4.2.2). A multidimensional box is constructed for each class and the classification of future samples is performed by determining within which box, if any, the sample belongs (using an F test).

Weaknesses

1. The statistical assumptions made for classification are often not obeyed by chemical data. Therefore, the F_{crit} values obtained from statistical tables are often not appropriate. For this reason, an empirical critical F value must be derived from validation and experience.

2. To obtain a good representation of the size and shape of a class, many samples per class must be obtained for deriving the SIMCA models.

Strengths

1. SIMCA has the ability to automatically detect if an unknown sample is not a member of any class in the training set.

2. SIMCA uses the shape and position of the object formed by the samples in a class for class definition.

3. SIMCA uses a PCA model for defining the class and, therefore, signal averaging can take place.

One of the differences between these two methods is in how prediction samples which do not belong to any known class are handled. SIMCA is inherently able to flag the new sample as "unusual," while KNN does not (without doing extra calculations as shown in Section 4.3.1). SIMCA has an additional advantage over KNN in that it uses the shape and position of the object formed by the samples in a class for class definition. Finally, SIMCA can include signal averaging because a PCA model is used for defining the classes. KNN has one advantage in that very few assumptions are made and is, therefore, the appropriate technique when there are few samples of known identity per class.

4.4 SUMMARY OF PATTERN RECOGNITION

Pattern recognition methods are used for elucidating or confirming groupings of samples in multivariate row space. In the unsupervised methods, the class identification(s) are not used in the calculation, and the goal is to examine the data for natural clusters without regard to assumed class membership. It is a good idea to always examine the data in this way regardless of the ultimate modeling goal. In the supervised methods, the known class membership is used to develop a model to classify samples in the future.

The two unsupervised methods examined are HCA and PCA. HCA calculates the interpoint distances between all of the rows and represents that information in the form of a two-dimensional plot called a dendrogram. PCA calculates a new axis system that maximally describes the variation in the data set. Our recommendation is to use both of the methods when they are available. HCA gives a broader view of the data and PCA can be used to further investigate samples and clusters that are highlighted in HCA.

Two supervised methods are examined in this chapter, KNN and SIMCA. The KNN models are constructed using the physical closeness of samples in space. It is a simple method that does not rely on many assumptions about the data. SIMCA models are based on more assumptions and define classes using the position and shape of the object formed by the samples in row space. A multidimensional box is constructed for each class using PCA and the classification of future samples is performed by determining within which box the sample belongs.

REFERENCES

W. P. Carey, K. R. Beebe, B. R. Kowalski, D. L. Illman, and T. Hirschfeld, *Anal. Chem.*, **58**, 149-153 (1986).

G. H. Golub and C. F. Van Loan, *Matrix Computations*, Johns Hopkins University Press, Baltimore, MD, 1983.

K. Hool, M. B. Seasholtz, R. Saunders, and D. Schlicker, *Incorporating Temporal Response Factors into Pattern Recognition Schemes for Gas Phase Sensor Arrays*, Paper No. 510, The Pittsburgh Conference, Chicago, IL, 1994.

E. R. Malinowski, *Factor Analysis in Chemistry*, 2nd, Wiley, New York, 1991.

H. Martens and T. Næs, *Multivariate Calibration*, Wiley, New York, 1989.

D. L. Massart, B. G. M. Vandeginste, S. N. Deming, Y. Michotte, and L. Kaufman, *Chemometrics: A Textbook*, Elsevier, New York, 1988.

5

Multivariate Calibration and Prediction

> As far as the laws of mathematics refer to reality, they are not certain; and as far as they are certain, they do not refer to reality.
>
> —*Albert Einstein, Sidelights on Relativity (1922)*

Calibration is the process of constructing a mathematical model to relate the output of an instrument to properties of samples. Prediction is the process of using the model to predict properties of a sample given an instrument output. For example, the absorbance at a given wavelength can be related to the concentration of an analyte. To construct the model, instrument responses from samples with known concentration levels are measured and a mathematical relationship is estimated which relates the instrument response to the concentration of the chemical component(s). This model may be used to predict the concentration of a chemical component in future samples using the measured instrument response(s) from those samples. Many instrumental responses can be considered, and a number of sample properties can be predicted. In this chapter, the terms spectrum and concentration are used generically to refer to the instrument response and sample property, respectively. This simplifies the discussion and is not meant to imply that the methods are limited to only spectroscopic applications for the determination of concentrations.

In many applications, one response from an instrument is related to the concentration of a single chemical component. This is referred to as univariate calibration because only one instrument response is used per sample. Multivariate calibration is the process of relating multiple responses from an instrument to a property or properties of a sample. The samples could be, for example, a mixture of chemical components in a process stream, and the goal is to predict the concentration levels of the different chemical components in the stream from infrared measurements. The methods are quite powerful, but as Dr. Einstein noted, the application of mathematics to reality is not without its limitations. It is, therefore, the obligation of the analyst to use them in a responsible manner.

Multivariate calibration offers several advantages over univariate approaches:

1. It is possible to analyze for multiple components simultaneously. This is not possible with univariate analyses and can reduce the amount of time spent on method development.

2. Multiple redundant measurements can also provide improved precision in prediction. Statistics show that repeating a measurement n times and calculating a mean value will result in a factor of $\sqrt{n}$ reduction in the standard deviation of the mean. This is commonly termed signal averaging. Analogous improvements can be gained by obtaining highly correlated measurements and applying the techniques described in this chapter. This is possible in most spectroscopic applications because adjacent measurements (wavelengths) are correlated.

3. Redundant measurements in multivariate calibration also facilitate fault-detection. Univariate analysis has no fault-detection capabilities, as illustrated in Figure 5.1 where the true relationship between the instrument response (on the y axis) and concentration (on the x axis) is represented by the solid line. It follows that a sample concentration of c_{true} corresponds to an instrument response of r_{true}. Errors occur if an unknown interferent (another component) is present and the instrument has a significant sensitivity to it. In Figure 5.1, assume $r_{true+interference}$ is the response of the instrument to a sample with concentration c_{true} plus some additional response due to an interferent. Using the calibration curve yields a predicted concentration, $\hat{c}$, for the analyte. The presence of

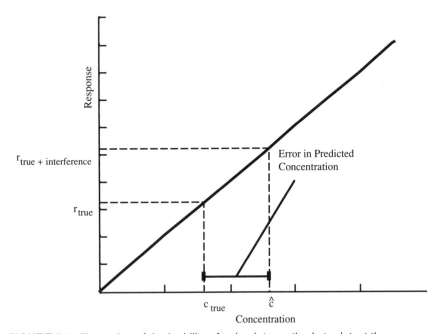

FIGURE 5.1. Illustration of the inability of univariate methods to detect the presence of interferents.

the interferent has resulted in the overestimation of the concentration of the analyte of interest as indicated in the figure. In addition, the analyst will not be aware of the problem. Using univariate measurement schemes, there are two ways to avoid interferent problems—physically separating the analytes of interest from all interferents and/or using selective measurements. Although both approaches are often acceptable and successful, considerable effort can be expended in developing the separations and/or measurements.

Multiple measurements allow the analyst to detect the presence of an interferent. For example, suppose an instrument outputs two responses for each sample, as shown in Figure 5.2. Assume the "pure component response" is obtained from the measurement of a pure analyte of interest. Given any sample containing only this analyte, the same relative magnitudes for measurements r_1 and r_2 are observed. If the observed response shown in Figure 5.2 is encountered, the analyst will know that a problem exists. This concept similarly extends to mixtures. The detection of interferents can fail if, for example, an interferent has the same relative response on the two variables as one of the other species. However, the likelihood of this type of failure decreases as the number of measurements increase, and as care is taken in choosing the measurements.

4. One of the more powerful features of multivariate calibration is that it can lead to paradigm shifts in problem solving. By allowing the analyst to consider nontraditional solutions to a problem, considerable savings and improved understanding are possible. An example of a paradigm shift is the use of NIR for quantitative analysis (Wetzel, 1983). For many years spectroscopists were unimpressed with this region of the electromagnetic spectrum because the signals overlapped. It is difficult to assign specific bands to specific chemical species, as is necessary when using traditional univariate approaches. More recently, chemometric techniques have been used in conjunction with the NIR region for

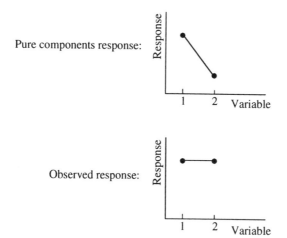

FIGURE 5.2. Illustration of the ability of multivariate methods to detect the presence of interferents.

quantitative analysis. The prediction of protein content in wheat, replacing the time-consuming and hazardous Kjehldahl method (Kjehldahl, 1883) and the determination of the octane number of gasoline (Kelly et. al., 1989) are two examples of a paradigm shift in solving important analysis problems.

5.1 DECISION TREE

The decision tree in Figure 5.3 is a guide for the selection of the appropriate multivariate calibration tool for a particular problem. Before discussing this tree, it is important to understand the distinction between linear and nonlinear approaches. In this chapter, only methods that estimate models that are linear in the parameters are discussed. Fortunately, these methods also tolerate some degree of nonlinear behavior in the data. When the nonlinearity cannot be modeled, some of the diagnostics (e.g., actual versus predicted value plots) may reveal nonlinearity as the problem, but this will not always be the case. In general, the failure of the methods is manifested in an inability to find a model with adequate performance.

In Figure 5.3 there are two main branches for the multivariate methods: classical least squares (CLS) and inverse least squares (ILS). Deciding which method to employ is not always straightforward. The question posed is "Is the system simple and are all of the analytes known?" If the answer to this question is yes, CLS *may* be the best approach. If the answer is no, ILS is the best (or only) choice. The difficulty in applying the CLS approach is knowing if the system meets the assumptions inherent to the methods. Classical least squares is applicable when all of the analytes are known and pure spectra can be obtained using either the direct (Section 5.2.1) or indirect (Section 5.2.2) methods. These systems typically have relatively few major components, few factors that affect the instrument response other than the concentration of the analytes of interest, and no significant interferences. It also must be possible to represent the mixture spectra as combinations of the pure spectra. If all of these conditions hold, CLS has an advantage over the ILS methods in that it is generally easier to construct a calibration model. All that is required for the model are spectra of the pure components of interest.

As indicated by the decision tree, two CLS methods are available, direct (DCLS) and indirect (ICLS), where the difference is in how the pure spectra are obtained. If it is possible to isolate the pure analytes and directly measure the pure spectra, the DCLS method can be used. If the pure analytes cannot be isolated, it may be possible to obtain the pure spectra mathematically using spectra of mixtures and the ICLS approach.

If the system is not simple, an inverse calibration method can be employed where it is not necessary to obtain the spectra of the pure analytes. The three inverse methods discussed later in this chapter include: multiple linear regression (MLR), principal components regression (PCR), and partial least squares (PLS). When using MLR on data sets found in chemistry, variable selection is

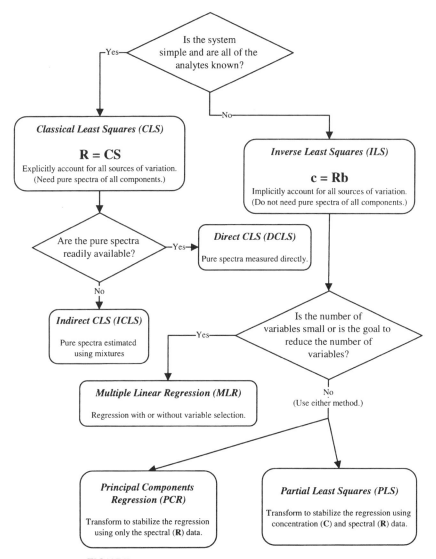

FIGURE 5.3. Multivariate calibration decision tree.

often required to make the matrix calculations possible and/or improve the stability of the calculations. It is therefore appropriate to use MLR when the number of variables is small, or in situations where a subset of measurement variables is desired. An example of the latter is when a full-wavelength instrument is used to perform a feasibility study, but where a filter instrument is to be purchased for economic or other reasons. Keep in mind that reducing the number of variables will almost always result in poorer error detection ability and less precise estimates. However, this may be outweighed by other factors.

Unlike MLR, PCR and PLS are methods that can be used without explicitly selecting variables. This is accomplished by transforming the measured variables (e.g., absorbance values at many wavelengths) into new variables (often referred to as factors) that are used in the matrix calculations. The difference between PCR and PLS is in how this variable transformation is performed. Both PCR and PLS have good diagnostic tools and in general the results are similar. These methods are often preferred over MLR unless the number of variables is small or circumstances dictate the explicit reduction in the number of variables.

The basic questions that must be answered in deciding between these methods are found in the decision tree in Figure 5.3. The discussions and information found throughout this chapter also help reveal how these questions are best answered.

5.2 CLASSICAL METHODS

The methods discussed in this section are termed classical least squares (CLS) in Figure 5.3. They can be used when the system under investigation obeys a linear relationship between the measurement vectors and concentration. An example of this type of relationship in spectroscopic applications is termed Beer's Law, which states that the spectral absorbance at a given wavelength is proportional to the concentration of a chemical component (Peters, 1974). For a single wavelength and a single chemical species, Beer's law is mathematically shown in Equation 5.1:

$$r_A = c_A a_A b_A \tag{5.1}$$

where r_A is the spectral response (in absorbance units) of the instrument to analyte A, c_A is the concentration (mole/L) of the analyte, b_A is the pathlength (cm), and a_A is the molar absorptivity (absorbance-L)/(mole-cm). In the equations below the pathlength is assumed to be constant and s_A (sensitivity) is substituted for the product of a_A and b_A.

Assuming Beer's Law holds for multiple wavelengths, $r_{A1} = c_A s_{A1}$, $r_{A2} = c_A s_{A2}$, . . ., $r_{Aj} = c_A s_{Aj}$, where r_{Aj} and s_{Aj} are the response and sensitivity, respectively, of the instrument to analyte A at wavelength j. For a multivariate measurement system, Equation 5.1 can therefore be expanded to

$$[r_{A1}\ r_{A2}\ r_{A3}\ \cdots\ r_{Aj}] = c_A\ [s_{A1}\ s_{A2}\ s_{A3}\ \cdots\ s_{Aj}] \tag{5.2}$$

and is represented in matrix form in Equation 5.3.

$$\mathbf{r} = c * \mathbf{s} \tag{5.3}$$

Linear *additivity* is also assumed when using the CLS models. That is, the response of an instrument to a mixture of analytes is equal to the sum of the instrument responses to the pure component responses. This holds in situations where there are no chemical or physical interactions. For wavelength j, linear additivity is expressed mathematically as

$$r_{ABj} = r_{Aj} + r_{Bj} = c_A s_{Aj} + c_B s_{Bj} \tag{5.4}$$

where r_{ABj} is the response of the instrument to a mixture of A and B. For all j wavelengths, this equation can be written as

$$[r_{AB1} \; r_{AB2} \; \cdots \; r_{ABj}] = [c_A \;\; c_B] \begin{bmatrix} s_{A1} \; s_{A2} \; \cdots \; s_{Aj} \\ s_{B1} \; s_{B2} \; \cdots \; s_{Bj} \end{bmatrix} \tag{5.5}$$

and in matrix notation as

$$\mathbf{r} = \mathbf{c}^*\mathbf{S} \tag{5.6}$$

The mixture spectrum on the left side of Equation 5.5 is said to be equal to a linear combination of the pure spectra on the right side of the equation. The top row of the $\mathbf{S}$ matrix is the spectrum of pure analyte A and the bottom row is the spectrum of pure B at unit concentration. The concentrations of the analytes are the coefficients for this linear combination. Figure 5.4 is a graphical illustration of Equation 5.6 where the mixture spectrum $\mathbf{r}$ is shown as a linear combination of three pure component spectra. The sample is made up of 1.8, 0.4, and 0.7 units of the analytes found in rows one, two, and three of the $\mathbf{S}$ matrix, respectively.

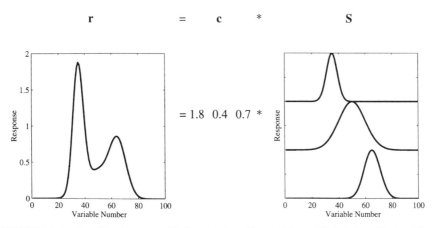

FIGURE 5.4. Graphical representation of Equations 5.5 and 5.6. The offset of the rows of **S** was added for clarity.

To construct a CLS model, the pure spectra for all analytes are obtained to form the **S** matrix. Two CLS methods are discussed, direct (Section 5.2.1), where the spectra are measured directly, and indirect (Section 5.2.2), where the pure spectra are computed from spectra of mixtures with known composition.

To perform prediction, an unknown mixture spectrum is measured (**r**). Given **r** and **S**, it is possible to solve Equation 5.6 for **c** (the predicted concentrations) using standard linear algebra:

$$\mathbf{r} = \mathbf{cS} \tag{5.7}$$

$$\mathbf{rS}^T = \mathbf{cSS}^T \tag{5.8}$$

$$\mathbf{rS}^T(\mathbf{SS}^T)^{-1} = \mathbf{c}(\mathbf{SS}^T)(\mathbf{SS}^T)^{-1} \tag{5.9}$$

$$\mathbf{rS}^T(\mathbf{SS}^T)^{-1} = \hat{\mathbf{c}} \tag{5.10}$$

where $\hat{\mathbf{c}}$ contains the estimated concentrations, and the superscripts T and -1 denote the transpose and inverse of a matrix, respectively. Equation 5.10 can also be written as:

$$\mathbf{rS}^\dagger = \hat{\mathbf{c}} \tag{5.11}$$

where $\mathbf{S}^\dagger$ is set equal to $\mathbf{S}^T(\mathbf{SS}^T)^{-1}$ and is called the pseudo-inverse of **S**.

The matrix $\mathbf{S}^\dagger$ can be used to simultaneously predict the concentrations of all analytes in one or multiple samples. Figure 5.5 shows the prediction of five unknown samples which contain the same three components shown in Figure 5.4. The $\mathbf{S}^\dagger$ in Figure 5.5 was calculated from the **S** matrix in Figure 5.4.

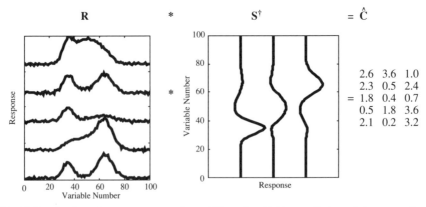

FIGURE 5.5. Graphical representation of Equation 5.11, prediction of five samples. The offset of the rows of **R** and $\mathbf{S}^\dagger$ was added for clarity.

Given the estimated concentrations, measurement residuals can be calculated for each sample and are examined for diagnostic purposes. The measurement residual is the portion of the mixture spectrum, $\mathbf{r}$, that is not described using the pure spectra. To generate a residual spectrum, a reconstructed mixture spectrum, $\hat{\mathbf{r}}$, is first generated using the pure spectra ($\mathbf{S}$) and the estimated concentrations ($\hat{\mathbf{c}}$):

$$\hat{\mathbf{r}} = \hat{\mathbf{c}}\mathbf{S} \tag{5.12}$$

This is what the mixture spectrum is expected to look like if the CLS assumptions are obeyed and $\hat{\mathbf{c}}$ is a reasonable estimate of the true concentrations. The residual is the difference between the observed and reconstructed mixture spectra.

$$\textbf{Residual} = \mathbf{r} - \hat{\mathbf{r}} = \mathbf{r} - \hat{\mathbf{c}}\mathbf{S} \tag{5.13}$$

The residual calculation is seen graphically in Figures 5.6 and 5.7 for sample number two (the second row of $\mathbf{R}$) in Figure 5.5. In Figure 5.6, the reconstructed spectrum is calculated using the known pure spectra and the predicted concentrations. In Figure 5.7, the measurement residual is obtained by subtracting $\hat{\mathbf{r}}$ from the observed $\mathbf{r}$.

Given that all three assumptions hold (linearity, linear additivity, and all pure spectra known), CLS has an advantage over the inverse methods (see Section 5.3) in that the calibration models are often easier to determine. For a simple system with three components, calibration may be as simple as obtaining the spectra of the three pure components.

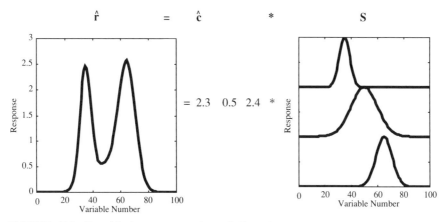

FIGURE 5.6. Graphical representation of Equation 5.12, estimation of mixture spectrum, $\hat{\mathbf{r}}$. The offset of the rows of $\mathbf{S}$ was added for clarity.

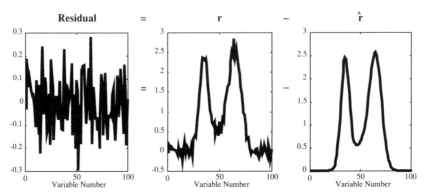

FIGURE 5.7. Graphical representation of Equation 5.13, estimation of the residual spectrum.

5.2.1 Direct CLS

In direct CLS (DCLS), the **S** matrix is obtained by measuring the spectra of the pure components (measured neat or dissolved in a nonabsorbing solvent). The details of this approach can be found in the introduction to this section. The following section describes how the method is applied to a data set.

5.2.1.1 DCLS Example 1   The example discussed here involves samples that contain two components, A and B. The pure spectra are obtained to construct the calibration model, and a test set of mixtures is used to validate the model. This test set is comprised of 26 samples where the concentrations of components A and B vary throughout the expected range of future samples. For this example, the reference value concentrations in the validation samples are determined using independent analytical methods. These values are not necessary to construct the model, but are used to validate the prediction performance. From Equation 5.11, the prediction model ($S^{\dagger}$) constructed from the measured pure spectra is used to predict the concentrations of the validation samples ($\hat{c}$) using the measured spectra of these samples (r). The predicted concentrations are then compared to the reference value concentrations to help validate the performance of the model. Often the goal of the study is to replace the independent analytical methods used to determine the concentrations of the validation samples with the new measurement and the DCLS model. This is advantageous when the independent method is difficult and/or time intensive.

Once the data are obtained, the model can be validated following the "Six Habits of an Effective Chemometrican" which are detailed in Chapter 1.

Habit 1. Examine the Data
Examine plots of the data for obvious violations of the CLS assumptions of linearity, linear additivity, and all pure component spectra known. The pure

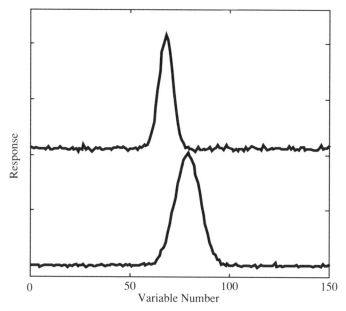

FIGURE 5.8. Pure spectra for Example 1—A (top) and B (bottom).

spectra (measured neat) are plotted in Figure 5.8. There is some overlap in the spectra and reasonable signal to noise ratio. The validation spectra shown in Figure 5.9 also look reasonable. The features present in these spectra are expected given the known concentrations and the pure spectra.

Examining the **C** matrix helps determine if the validation data are reasonable for validating the CLS model. A good distribution of concentrations spanning the ranges of interest is required. Figure 5.10 is a plot of the concentration of component B versus the concentration of component A for the validation samples. This plot verifies that the validation data have a reasonably good distribution. The samples were not collected according to a formal experimental design, but according to a "natural design" (see Martens and Næs, 1989, and Appendix A of Chapter 2).

Habit 2. Preprocess as Needed

If the pure component samples were prepared in a transparent matrix, the spectra should be normalized to produce a pure spectrum at unit concentration. It is also common to estimate an offset or randomly varying linear or quadratic baseline when using DCLS. To estimate an offset, a vector of ones is added to the **S** matrix as another component. For a linear baseline, a running index vector and a vector of ones are added to the **S** matrix. The multiplier for the running index accounts for the slope of the baseline and the vector of ones accounts for the offset. Finally, for a quadratic baseline, a running index squared vector, a running index vector, and a vector of ones are added to the

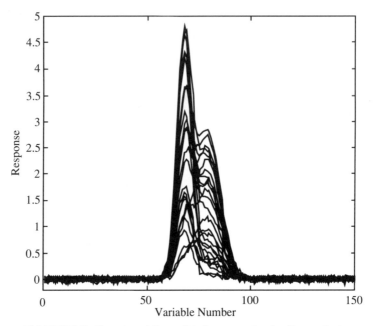

FIGURE 5.9. Spectra of the validation samples for Example 1.

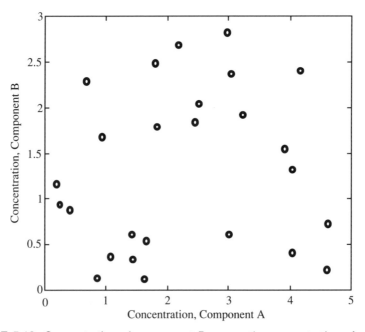

FIGURE 5.10. Concentration of component B versus the concentration of component A for the validation samples.

S matrix. Higher-order baseline models are implemented by appending running index vectors with increasing exponents to the **S** matrix. However, using polynomials higher than second order can lead to matrix inversion problems. For this example, no features are present (e.g., baseline, spikes, or large noise) that indicate the need for further preprocessing.

Habit 3. Estimate the Model

To estimate the model ($\mathbf{S}^\dagger$), the pure component spectra (with any augmentations for baseline) are supplied to the computer. For validation, the **R** and **C** matrices are submitted. The resulting output includes the statistical prediction errors, estimated concentrations, concentration residuals, and spectral residuals. For this example, no baselines are estimated; only the pure component spectra are used for **S**.

Habit 4. Examine the Results/Validate the Model

Several diagnostic tools are discussed below and a summary is found at the end of the section in Table 5.1. These tools are used to investigate three aspects of the data set: the model, the samples, and the variables. The headings for each tool indicate the aspects that are studied with that tool. The primary use of the model diagnostic tools is to investigate whether the CLS assumptions hold. The sample diagnostic tools are used to identify unusual samples. Finally, the variable diagnostics are used to identify abnormal variables within a spectrum that may indicate instrumental problems.

Statistical Prediction Errors (Model and Sample Diagnostic): Uncertainties in the concentrations can be estimated because the predicted concentrations are regression coefficients from a linear regression (see Equations 5.7–5.10). These are referred to as statistical prediction errors to distinguish them from simple concentration residuals ($c - \hat{c}$). The statistical prediction errors are calculated for one prediction sample as

$$\text{sd}(\mathbf{c}) = \sqrt{\text{diag}(\mathbf{SS}^T)^{-1} * s^2} \tag{5.14}$$

where sd(**c**) is a vector containing the statistical prediction errors for the different components (one standard deviation), **S** is the matrix of pure spectra, $\text{diag}(\mathbf{SS}^T)^{-1}$ is a vector containing the diagonal elements of $(\mathbf{SS}^T)^{-1}$, and s^2 is a scalar value which is a measure of the portion of the unknown spectrum that is not described by the pure spectra (spectral residuals). In statistics, s^2 is termed the mean square about the regression (Draper and Smith, 1981). (Equation 5.14 does not look exactly like the equations found in statistics books because the **S** matrix is oriented with the pure spectra in the rows to conform with the notation in this book.)

The statistical prediction error is in concentration units and represents the uncertainty in the predicted concentrations due to deviations from the model assumptions, measurement noise, and degree of overlap of the pure spectra. As the system deviates from the underlying assumptions of CLS, the residual spectrum will have larger features (increase in s^2). If the measurement noise is high, s^2 will also increase. If the pure spectra are very correlated (i.e., highly overlapped), the statistical prediction errors will increase because the elements in the $(SS^T)^{-1}$ matrix become larger.

The statistical prediction error does not account for biases in concentration or pathlength changes. The $(SS^T)^{-1}$ matrix depends only on the pure spectra, and the residual spectra only depends on how well CLS can find a linear combination of pures to fit the sample spectrum. In other words, the statistical prediction error is a measure of precision, not accuracy.

If the statistical prediction errors for the test set are not acceptable for the targeted application, either the model assumptions are not met, the measurement noise is too high, or the pure spectra are too similar. One approach for increasing the difference between the pure spectra is to eliminate measurement variables that are highly correlated between the pure spectra. Removing regions with no discriminating power results in smaller diagonal values in the $(SS^T)^{-1}$ matrix and correspondingly smaller statistical prediction errors.

The statistical prediction errors for the validation samples in this example are plotted in Figure 5.11. The maximum statistical prediction error for component A (Figure 5.11a) is ~0.025, which must be compared to the precision requirements of the application. For component B (Figure 5.11b), maximum statistical prediction error is ~0.019. In addition, no samples appear to have unusually large statistical prediction errors which would indicate an outlier.

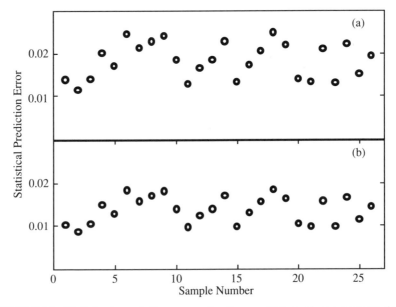

FIGURE 5.11. Statistical prediction errors for the prediction of components (a) A and (b) B.

Predicted vs. Known Concentration Plot (Model and Sample Diagnostic):
The next plot to examine is the concentration for each validation sample predicted using the DCLS model versus the known concentration derived from the reference method. Figure 5.12 shows three behaviors that can be observed when examining this plot. The ideal behavior is to have the points follow a line of slope 1 and intercept 0 as illustrated in Figure 5.12*a*. If this is not the case, the extent to which the intercept is nonzero indicates a constant bias between the known and predicted values. A slope not equal to 1 indicates a relative bias. Structured patterns in this plot also indicate problems (see Figure 5.12*b*). Nonideal intercepts or slopes, and/or patterns in the plot all indicate problems with the model. The pattern shown in Figure 5.12*c* indicates the presence of an outlier, that is, a sample that has a quite different predicted concentration than expected based on the behavior of other samples in the validation set.

This plot can help point out a problem, but finding the cause often takes chemical knowledge. Figure 5.12*c* indicates the actual concentration and/or the predicted concentration is in error for one sample. The first step is to examine the outlier measurement vector relative to the other samples and use human pattern recognition to propose root causes. If nothing unusual is found in the measurements, the next step is to consider the actual concentration values. Are the true concentration values consistent with the measurements? An inconsistency may indicate a transcription error or that the sample was prepared incorrectly. While the root cause of a problem may seem obvious from interpretation of this one diagnostic, we recommend a complete examination of all the diagnostic output before settling on any conclusions.

The predicted versus known concentration plot for component A for this example is shown in Figure 5.13*a*. The data appear to be randomly distributed about the ideal line, indicating the model is correct. One sample appears to be unusual, which warrants further investigation. The same graph for component B is shown in Figure 5.13*b*. As with component A, the data appear to be randomly distributed about the ideal line. None of the samples appear to be unusual when predicting component B.

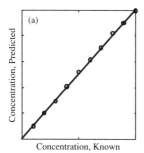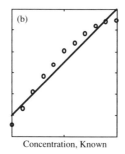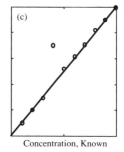

FIGURE 5.12. Possible behavior of actual versus predicted concentration plots: (*a*) ideal, (*b*) nonlinear, and (*c*) outlier.

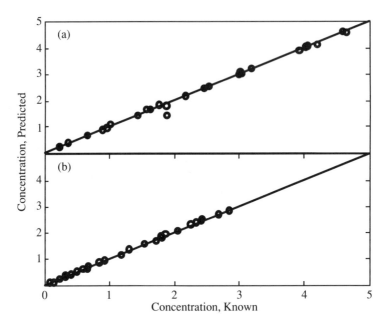

FIGURE 5.13. Predicted concentration versus known concentration for component A (*a*) and B (*b*). The solid line has a slope of 1 and an intercept of 0.

Concentration Residuals vs. Predicted Concentration Plot (Model and Sample Diagnostic): The plot of concentration residuals ($c - \hat{c}$) versus predicted concentration gives similar information to the predicted vs. known plots. Ideally, this plot has a scatter of points about a line with slope and intercept equal to zero. Figure 5.14 shows the concentration residual versus predicted plots that correspond to the scenarios presented in Figure 5.12. The residual plot enhances features that can go unnoticed in the actual versus predicted plots.

The concentration residuals are plotted as a function of the predicted concentration (Draper and Smith, 1981). One side note is that with replicate samples it can appear that there is structure in the residuals when in fact the model is adequate. This is demonstrated in Figure 5.15 where there are 10 replicates at three concentration levels. The residuals form slanted lines, but this is only because some samples are predicted low (and so have positive errors), and other samples are predicted high (and so have negative errors).

Returning to the example, the concentration residuals versus predicted concentration are plotted in Figure 5.16. The unusual sample for component A is very apparent with a residual of 0.46 while the rest of the residuals are within ±0.1. Except for this single sample with large residual, the residuals appear randomly distributed about zero, indicating an adequate model. No samples appear to be unusual from the residuals for component B (Figure 5.16*b*).

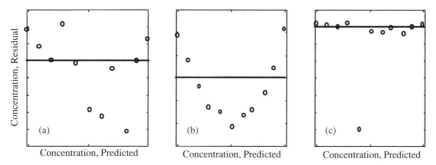

FIGURE 5.14. Residual versus predicted concentration plots for data found in Figure 5.12: (a) ideal, (b) nonlinear, and (c) outlier.

The conclusion is that the model appears to be acceptable. This graph also provides information about how well the method will predict future samples. It is expected that the errors in prediction for component B will be ±0.06. This conclusion is only possible because the validation set contains many samples that adequately span the calibration space (see Habit 1). A conclusion about the prediction errors for component A will be evaluated after resolving the issue with the unusual sample.

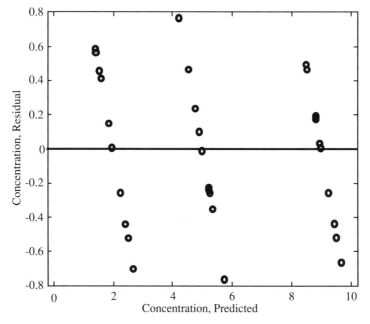

FIGURE 5.15. Concentration residual versus predicted concentration for the case of replicate analyses.

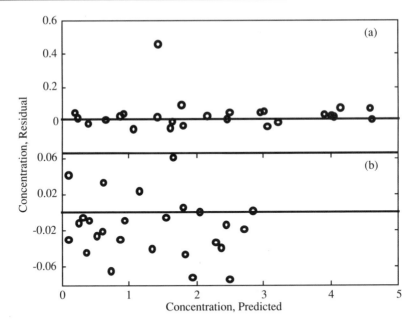

FIGURE 5.16. Concentration residual versus predicted concentration for (a) component A and (b) component B.

Concentration Residuals vs. Other Postulated Variable Plot (Model and Sample Diagnostic): It can be helpful to plot the concentration residuals versus another parameter that is postulated to be important. For example, it may be suspected that the order in which the samples were run has an effect on the concentration errors due to changes in the room temperature. Without the a priori information or hypothesis, these dependencies can go undiscovered, because the patterns appear only when the x axis represents a variable that is affecting predictions.

To investigate whether run order influenced the results for this example, the component A concentration residuals are plotted versus the order in which the reference values for component A were measured (Figure 5.17). The first sample is the one with the unusually high error, which may indicate problems with the startup of the instrument used to determine the reference concentration of component A. What appears to be a nonrandom pattern over time may indicate instrumental fluctuations in the reference method determination.

Root Mean Square Error of Prediction (RMSEP) (Model Diagnostic): The RMSEP is another diagnostic for examining the errors in the predicted concentrations. While the statistical prediction error discussed earlier quantifies preci-

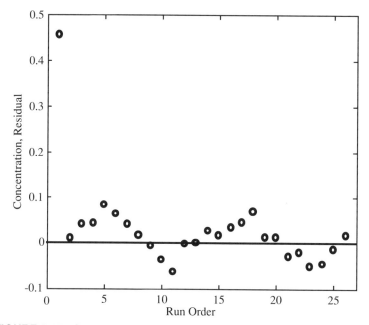

FIGURE 5.17. Concentration residuals versus run order for component A.

sion, the RMSEP shown in Equation 5.15 summarizes both the precision and accuracy of future predictions:

$$\text{RMSEP} = \sqrt{\frac{\sum_{i=1}^{\text{nsamp}} (c_i - \hat{c}_i)^2}{\text{nsamp}}} \qquad (5.15)$$

where c_i is the true concentration of the component of interest in the ith sample of the validation set, $\hat{c}_i$ is the estimated concentration using DCLS, and nsamp is the number of samples in the validation set. In statistics, the RMSEP is referred to as a root mean square error. The RMSEP summarizes the spread of the concentration errors into one number similar to a standard deviation and in the same units as the concentration values.

The RMSEP is 0.06 and 0.04 for components A and B, respectively. The RMSEP for component A calculated without the unusual sample is 0.04. The range of concentration errors observed for component B in Figure 5.16 is ±0.06, which is consistent with a RMSEP of 0.04 (the full range of concentration residuals should correspond to approximately 2–3 RMSEP units if there is no bias).

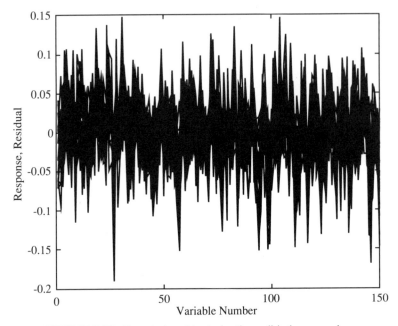

FIGURE 5.18. Spectral residuals for the validation samples.

Measurement Residual Plot (Model, Sample, and Variable Diagnostic):
The measurement residual, $\hat{\mathbf{r}}$, is the portion of the measured vector that is not
fit by the model. It is generated using the measured vector, $\mathbf{r}$, the estimated
concentrations, $\hat{\mathbf{c}}$, and the pure component matrix, $\mathbf{S}$ (see Equations 5.12 and
5.13). If the model is appropriate and $\hat{\mathbf{c}}$ is a good estimate of the true concentrations, the residuals will have random variation corresponding in magnitude
to the instrumental noise. Human pattern recognition is used to decide
whether the plot is exhibiting reasonable behavior.

The spectral residuals for the validation samples are plotted in Figure 5.18.
For all samples, the residuals appear to be randomly varying between ± 0.15
units which is within the expected instrumental noise.

Summary of Validation Diagnostic Tools for DCLS, Example 1: The spectra from two pure components were measured and without further correction
were used in a two-component model. The details of the data set used to construct the calibration model are as follows:

Calibration design: 2 pure-component spectra

Validation design: 26 samples, randomly varying in concentration

Preprocessing: none

Variable range: 100 original measurement variables

The operating ranges of the prediction model based on the validation data follow. Predicting future samples from outside this operating range is extrapolating and may produce unreliable results.

A: 0.2–4.6

B: 0.11–2.8

The conclusion from the validation of Example 1 is that the CLS model assumptions are valid. The measures of performance for the DCLS model are as follows (see Table 5.1 for a description of these figures of merit):

RMSEP = 0.04

Maximum statistical prediction error:

    0.025 for Component A

    0.019 for Component B

Expected error in prediction for components A and B = ± 0.06

(Based on errors in predicted concentration. See Figure 5.16.)

For one sample the predicted concentration for component A was very different from the reported value. For this sample, the prediction of the level of component B was reasonable and the spectral residuals were similar to the other validation samples. The similarity of the spectral residuals indicates the sample did not have unusual spectral features. The conclusion is that the reported or known concentration of component A is questionable. Recall that this sample was the first one analyzed when the reference value for the concentration of component A was determined (Figure 5.17) and time-related reference method problems were postulated. This should be confirmed by reanalyzing the sample.

Table 5.1 summarizes the validation diagnostic tools discussed in this section. The first column in the table lists the name of each tool and the second column describes results from both well-behaved and problematic data.

Habit 5. Use the Model for Prediction

The prediction step is the multiplication of the unknown spectrum ($\mathbf{r}$) by $\mathbf{S}^\dagger$, (see Equation 5.11). Remember to first apply any preprocessing to $\mathbf{r}$ before the multiplication by $\mathbf{S}^\dagger$; in this case, no preprocessing was used. The predicted concentrations of components A and B for four unknowns are shown in Table 5.2.

Habit 6. Validate the Prediction

One of the most powerful advantages of multivariate calibration is that the predictions can be validated. That is, it is possible to evaluate the reliability of the concentration estimates. Four prediction diagnostic tools are discussed below and a summary is found at the end of the section in Table 5.4.

TABLE 5.1. Summary of Validation Diagnostics for DCLS

Diagnostics	Description and Use
Model	
Statistical prediction errors (statistical prediction error vs. sample number)	The statistical prediction errors are the uncertainty in the predicted concentrations which reflects the quality of the model.
	They are an optimistic estimate of prediction ability (they only reflect precision) and should be evaluated relative to the requirements of the application.
	They are small or comparable to the known concentration errors when the model is performing well.
Predicted vs. known concentration plot ($\hat{c}$ vs. c)	Predicted concentrations of validation samples plotted versus the known values.
	Used to determine whether the model accounts for the concentration variation in the validation set.
	Points are expected to fall on a straight line with a slope of one and a zero intercept.
	A modeling problem is indicated when a systematic pattern away from the ideal line is observed.
Concentration residual vs. predicted concentration plot [$(c - \hat{c})$ vs. $\hat{c}$]	The differences between the known and predicted concentrations (residuals) are plotted versus the predicted concentrations for validation samples.
	Used to determine whether the model accounts for the concentration variation in the validation set.
	Points are expected to vary randomly about a line of zero slope and zero intercept.
	A modeling problem is indicated when a systematic pattern away from the ideal line is observed.
	Patterns are often more discernible in this plot compared to the predicted versus known plot.
Concentration residual vs. other postulated variable plot [$(c - \hat{c})$ vs. x]	The concentration residuals can be plotted versus a postulated variable such as run order.
	Used to determine if the model is accounting for the variability correlated with the postulated variable.
	Points are expected to vary randomly about a line of zero slope and zero intercept.
	A modeling problem is indicated when a systematic pattern away from the ideal line is observed.
	Identified effects should be controlled or added to the model if possible.
Root mean square error of prediction (RMSEP)	A single value that quantifies the magnitude of the concentration residuals for the validation samples.
	The RMSEP characterizes both the accuracy and precision errors expected for future predictions.
	When the model is performing well, the RMSEP should be comparable to the known error in the concentration reference method.
Measurement residual plot [$(r - \hat{r})$ vs. variable number]	The residuals are the portion of the sample measurement vector that is not fit by the pure spectra.
	Used to identify inadequacies in the model.
	Residual values are expected to vary randomly about a line of zero slope and zero intercept.

TABLE 5.1. *(Continued)*

Diagnostics	Description and Use
	Model error is suspected if there is a pattern for a number of samples away from the ideal line.
Sample	
Statistical prediction errors (statistical prediction error vs. sample number)	The statistical prediction errors are the uncertainty in the predicted concentrations for each validation sample.
	It is an optimistic estimate of the concentration error for each sample because it only includes precision.
	Statistical prediction errors for a sample that are quite different than the other validation samples indicate a problem with that sample.
Predicted vs. known concentration plot ($\hat{c}$ vs. c)	Predicted concentrations of validation samples plotted versus the known values.
	This plot can be used to identify outlier samples.
	Points are expected to fall on a straight line with a slope of one and a zero intercept.
	An outlier is indicated when a point is far from the ideal line relative to the other points.
Concentration residual vs. predicted concentration plot [$(c - \hat{c})$ vs. $\hat{c}$]	The differences between the known and predicted concentrations (residuals) are plotted versus the predicted concentrations for the validation samples.
	This plot can be used to identify outlier samples.
	Points are expected to vary randomly about a line of zero slope and zero intercept.
	A small number of samples with unusual concentration residuals is an indication of sample rather than model problems.
	Outliers are often more discernible in this plot compared to the predicted versus known plot.
Measurement residual plot [$(r - \hat{r})$ vs. variable number]	The residuals are the portion of the sample measurement vector that is not fit by the pure spectra.
	Used to identify outlying samples.
	Residual values are expected to vary randomly about a line of zero slope and zero intercept.
	An outlier is identified as a sample with a residual vector that is significantly different from the other residual vectors.
	Examining the variables where a sample has unusual features can be used to determine why the sample is unusual.
Variable	
Measurement residual plot [$(r - \hat{r})$ vs. variable number]	The residuals are the portion of the sample measurement vector that is not fit by the pure spectra.
	Used to identify outlying variables.
	Residual values are expected to vary randomly about a line of zero slope and zero intercept.
	Problematic variables are identified as those with unusual residual features in all of the samples.

TABLE 5.2. Concentration Predictions for Four Unknown Samples

	Component A	Component B
	Estimated	Estimated
Unknown	Concentration	Concentration
1	4.24	2.52
2	4.23	2.53
3	3.97	3.20
4	4.31	2.47

Predicted Concentration Plot: The predicted concentrations for the four unknowns are plotted in Figure 5.19. Except for the predicted component B concentration for sample 3, all concentrations are within the range of the validation data (indicated by the horizontal lines). The assumptions of the model were only validated within this concentration range, and therefore the prediction for sample 3 is suspect.

Statistical Prediction Errors: The statistical prediction errors for the unknowns are shown in Table 5.3. These are an estimate of the precision of the predicted concentrations (Equation 5.14). These error estimates are different for the two components even though the same spectral residuals are used in the calculation (s^2 in Equation 5.14). This is because there are different elements on the diagonal of the $(\mathbf{SS}^T)^{-1}$ matrix. Component B has a smaller

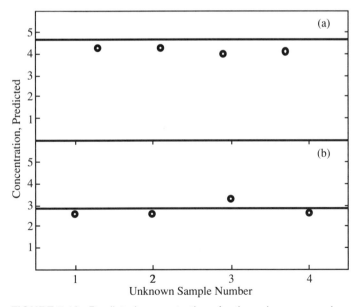

FIGURE 5.19. Predicted concentrations for the unknown samples.

**TABLE 5.3. Statistical Prediction Errors for
Four Unknowns**

	Component A	Component B
Unknown	Statistical Prediction Error	Statistical Prediction Error
1	0.025	0.019
2	0.174	0.129
3	0.157	0.117
4	0.128	0.095

diagonal term because it has more area per unit concentration and so can be more precisely determined than component A.

The statistical prediction errors for the unknowns are compared to the maximum statistical prediction error found from model validation in order to assess the reliability of the prediction. Prediction samples which have statistical prediction errors that are significantly larger than this criterion are investigated further. In the model validation, the maximum error observed for component A is 0.025 (Figure 5.11a) and 0.019 for component B (Figure 5.11b). For unknown 1, the statistical prediction errors are within this range. For the other unknowns, the statistical prediction errors are much larger. Therefore, the predicted concentrations should not be considered valid.

Measurement Residual Plot: The statistical prediction errors indicate which samples have large spectral residuals. It can be instructive to then plot the residuals to diagnose the problem. In practice, only samples with large statistical prediction errors are examined, but all four will be plotted here. The residuals for unknowns 1–4 are shown in Figures 5.20–5.23, respectively. Also shown are the measured and predicted responses. The residual for unknown 1 in Figure 5.20 resembles the model validation residuals shown in Figure 5.18.

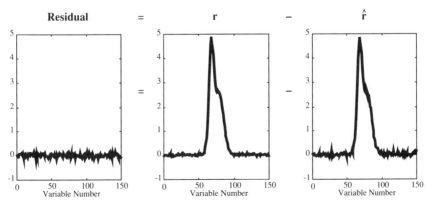

FIGURE 5.20. Raw measurement and spectral residuals for unknown 1.

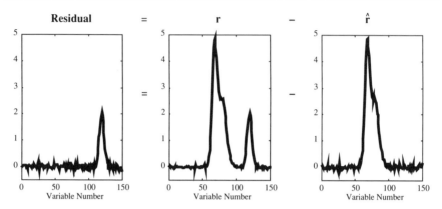

FIGURE 5.21. Raw measurement and spectral residuals for unknown 2.

This is expected given that the statistical prediction errors were comparable to those found from the validation samples.

For unknown 2, a peak centered at variable 120 is observed in the residual (Figure 5.21). The conclusion is that there is an interfering component with a pure spectrum that does not overlap with the pure spectra of A or B. Because it is not overlapped, the predicted concentrations are likely to be correct, even though the statistical prediction errors are large. The large statistical prediction errors in this case have indicated a problem which should be investigated. It may be that the variables used for the measurement vector at and around 120 should be removed.

The residual for unknown 3 has a feature centered around variable 90 (see Figure 5.22). The fact that it is in the same region as the response from the pures and has an unusual shape (it is unlike the shape of the pure spectra) suggests that an interferent is overlapped with components A and B. Unlike unknown 2, this cannot be detected by examining the raw data. The number of interferents, the spectral features, or the concentrations of the interfering

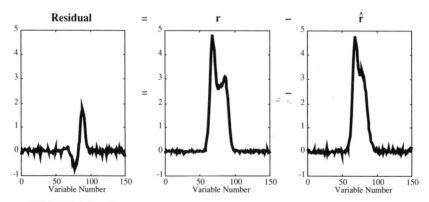

FIGURE 5.22. Raw measurement and spectral residuals for unknown 3.

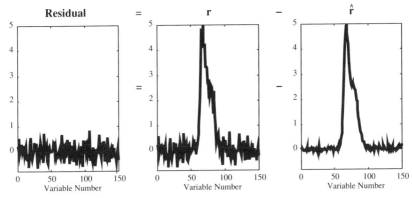

FIGURE 5.23. Raw measurement and spectral residuals for unknown 4.

species cannot be determined. Furthermore, the predicted concentrations of components A and B cannot be considered accurate.

The residuals for unknown 4 are predominantly random but have greater magnitude than expected (see Figure 5.23). This indicates that there are no interfering components in the sample but the noise level has increased. If this is the root cause, the predictions will in general be accurate, but will have poorer precision.

Raw Measurement Plot: The raw data for the four unknowns are plotted in Figures 5.20–5.23 and were discussed above.

Summary of Prediction Diagnostic Tools for DCLS, Example 1: The conclusions from the prediction diagnostics are that the predicted concentrations for unknowns 1 and 2 can be considered reliable. The predicted concentration of unknown 3 is unreliable, and no conclusive statement can be made about unknown 4.

Table 5.4 summarizes the prediction diagnostic tools discussed in this section. The first column in the table lists the name of each tool and the second column describes results from both well-behaved and problematic data.

"What If" Baseline Offset

Suppose that another series of samples is to be predicted using the DCLS model. Figure 5.24 displays the spectral residuals from the predictions. A problem is indicated because the residuals do not resemble those from the validation data (see Figure 5.18). Figure 5.25 displays the spectra of the unknown samples, which reveal a random linear baseline with variable offset. It is not obvious from the residuals that this is the problem, because CLS attempts to fit the baseline feature with the pure spectra. When unexpected features appear in the spectra of the prediction samples, CLS compensates by overestimating the concentration of one or more of the pure components. The result is that the residuals have features that come from the pure spectra as well as some remnant of the original unexpected features. This makes interpretation of the residual spectra difficult.

TABLE 5.4. Summary of Prediction Diagnostics for DCLS

Prediction Diagnostics	Description and Use
Predicted concentration plot (**c** vs. sample number)	A graphical means to display predicted concentrations along with the calibration range.
	Used to help assess the reliability of the predicted concentrations of unknown samples.
	The predicted concentrations should fall within the calibration concentration range.
Statistical prediction errors	The statistical prediction errors are the uncertainty in the predicted concentrations for each unknown sample.
	Used to help assess the reliability of the predicted concentrations of unknown samples.
	The predicted concentrations for an unknown are not reliable if the statistical prediction errors are significantly larger than observed for the calibration samples.
Measurement residual plot [(**r** − **r̂**) vs. variable number]	The residuals are the portion of the unknown sample measurement vector that is not fit by the pure spectra.
	Used to help assess the reliability of the predicted concentrations of unknown samples.
	Residual values are expected to vary randomly about a line of zero slope and zero intercept.
	The predicted concentrations are suspect if the residual vector is significantly different from the calibration residuals.
	Examining the variables where a sample has unusual features can be used to determine why the unknown sample is unusual.
Raw measurement plot [**r** vs. variable number]	A plot of the measurement vectors for the prediction samples.
	Used to investigate samples that have been identified as being unusual.
	Used in conjunction with knowledge of the chemistry and the measurement system to assign cause when suspect predictions are identified.

To correct the model, a random linear baseline is added by augmenting the **S** matrix with a vector of ones and a vector containing a running index. The resulting CLS model now yields four values, two for the concentrations of components A and B, and the other two for the baseline. Using the augmented **S** matrix, the resulting residuals (not shown) are similar to those found in Figure 5.18. In summary, the residuals are good for flagging problems, but do not always indicate the root cause.

5.2.1.2 DCLS Example 2   The data set for Example 2 consists of NIR spectra collected on mixtures of four organic liquids. The spectral intensities are measured from 1100 to 2500 nm on mixtures containing varying amounts of monochlorobenzene (MCB), ethylbenzene (EB), *o*-dichlorobenzene (ODCB),

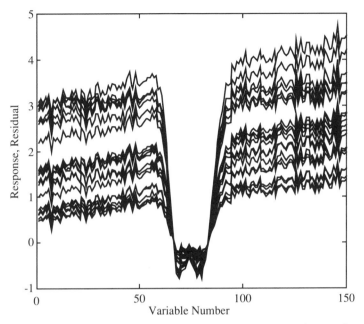

FIGURE 5.24. Spectral residuals from prediction of unknowns using a misspecified model.

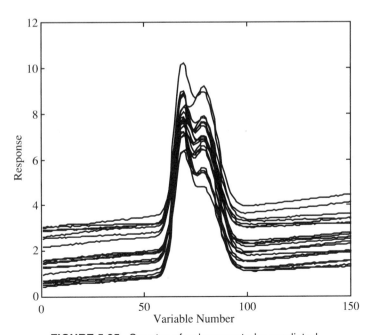

FIGURE 5.25. Spectra of unknowns to be predicted.

TABLE 5.5. Concentration Levels Used for DCLS Example 2

MCB	EB	ODCB	CUM
1.0000	0	0	0
0	1.0000	0	0
0	0	1.0000	0
0	0	0	1.0000
0	0	0.4999	0.5001
0	0.5000	0	0.5000
0	0.5000	0.5000	0
0.5000	0	0	0.5000
0.5000	0	0.5000	0
0.5000	0.5000	0	0
0	0.3332	0.3336	0.3332
0.3331	0	0.3327	0.3342
0.3317	0.3304	0	0.3380
0.3333	0.3333	0.3333	0
0.2500	0.2500	0.2498	0.2501
0.2494	0.2517	0.2494	0.2494
0.2501	0.2500	0.2499	0.2500
0.2499	0.2500	0.2499	0.2502
0.6239	0.1248	0.1250	0.1264
0.1229	0.6134	0.1226	0.1411
0.1249	0.1250	0.6247	0.1255
0.1291	0.1291	0.1288	0.6131

and cumene (CUM). A simplex centroid mixture design was used to choose the validation samples. In this design all four components vary from 0.00 to 0.62 weight fraction. There are 11 samples for the basic design (including one center point), four interior points, and three additional replicates of the center point. See Table 5.5 for the specific concentrations. The first four samples are the pure components used to construct the model and the remaining samples are used for model validation.

Habit 1. Examine the Data

The observed peaks in the four pure spectra shown in Figure 5.26 are consistent with the known chemistry and spectroscopy. Because linearity must hold for the CLS model to be valid, the regions with high absorbances will be eliminated in the preprocessing.

Figure 5.27 displays the spectra of the pure components along with the validation samples. No individual spectrum appears anomalous, but again high absorbances are observed along with a slight baseline offset.

Habit 2. Preprocess as Needed

Due to the high absorbances, the 2200- to 2500-nm region is deleted leaving the region from 1100 to 2198 nm for use in the model.

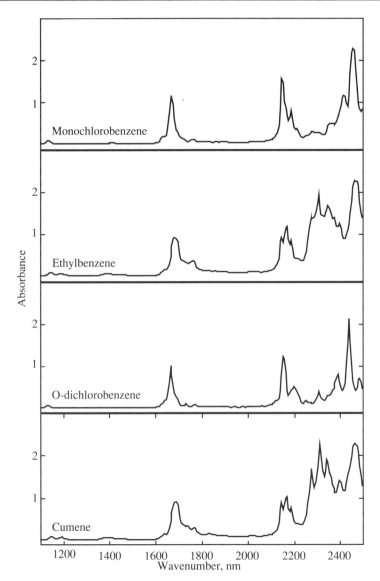

FIGURE 5.26. Near-infrared spectra of the four pure components for DLCS Example 2.

Habit 3. Estimate the Model

To estimate the model ($S^{\dagger}$), the 550 response variables in the region from 1100 to 2198 nm of the pure component spectra and a simple baseline offset are supplied to the computer. The resulting output from the validation samples includes the statistical prediction errors, estimated concentrations, concentration residuals, and spectral residuals.

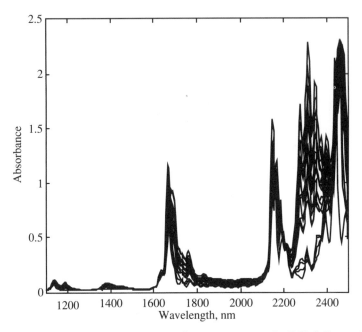

FIGURE 5.27. The pure component and test set spectra for DCLS Example 2.

Habit 4. Examine the Results/Validate the Model

To test whether the model is reasonable, the results from predicting the validation samples are examined.

Statistical Prediction Errors (Model and Sample Diagnostic): Figure 5.28 shows the statistical prediction errors for all four components for the samples in the validation set. For MCB and ODCB the maximum value is ~0.004 and for EB and CUM the maximum value is ~0.01. These errors are small compared to the concentration ranges.

One approach for determining whether the statistical prediction errors are reasonable is to compare them to statistical prediction errors computed using an estimate of s^2 from replicate analyses. Analysis of the replicate center point spectra provide an estimate of s^2 equal to 1.2e-07 AU. The diagonal elements of $(S^TS)^{-1}$ are 0.37, 4.9, 0.52, 5.7, and using Equation 5.14 to estimate statistical prediction errors for MCB, EB, ODCB, and CUM yields 0.0002, 0.0007, 0.0002, and 0.0008, respectively. These estimates based on replicate samples are an order of magnitude smaller than observed, indicating the model is inadequate.

Predicted vs. Known Concentration Plot (Model and Sample Diagnostic): The predicted versus known concentration plots for MCB and ODCB are shown in Figure 5.29a and b, respectively. The plots for EB and CUM (not shown) have a structure similar to ODCB. All predictions are biased as indicated by a systematic departure of the points from the ideal line. This indicates a modeling problem.

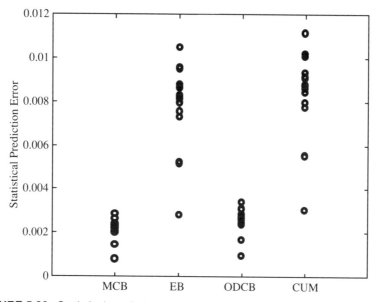

FIGURE 5.28. Statistical prediction errors for the four components in the mixtures.

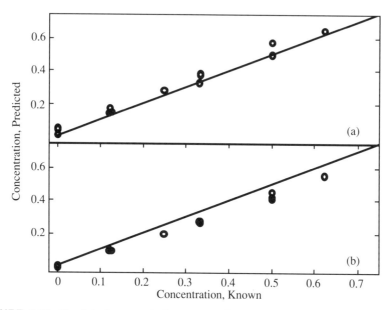

FIGURE 5.29. Predicted concentration versus known concentration for (*a*) mono-chlorobenzene and (*b*) *o*-dichlorobenzene.

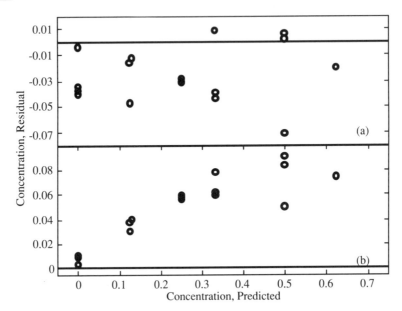

FIGURE 5.30. Concentration residual versus predicted concentration for (a) mono-chlorobenzene and (b) o-dichlorobenzene.

Concentration Residuals vs. Predicted Concentration Plot (Model and Sample Diagnostic): The bias in the predictions is more easily seen if the concentration residuals are plotted against the predicted values, as shown in Figure 5.30. For MCB (Figure 5.30a) the bias appears to be constant as a function of the predicted MCB, while for the ODCB (Figure 5.30b), the bias increases with increasing predicted ODCB.

Root Mean Square Error of Prediction (RMSEP) (Model Diagnostic): The RMSEP values for all four components are numerically summarized in Table 5.6. They are large owing to the bias in the predictions. Several reasons for this bias can be proposed, including: an inaccurate reference method, transcription errors, poor experimental procedures, changes in density and/or pathlength, light scatter in the instrument or sample, chemical interactions,

TABLE 5.6. Standard Error of Prediction for Four-Component Organic Mixture DCLS, Example 2

Component	RMSEP
Monochlorobenzene (MCB)	0.033
Ethylbenzene (EB)	0.033
o-Dichlorobenzene (ODCB)	0.055
Cumene (CUM)	0.049

and nonlinearity between instrument response and concentration. Determining the exact cause of the bias requires more experimentation. However, before conducting more experiments, it is wise to look at the residual spectra to see if they indicate the source of the problem.

Measurement Residuals Plot (Sample and Variable Diagnostic): There is nonrandom behavior in the spectral residuals, indicating inadequacies in the model (see Figure 5.31). This is consistent with the statistical prediction errors being an order of magnitude larger than the ideal value. Several preprocessing options were attempted without realizing improved prediction results. Without further investigation, it is not possible to determine the cause of the model failure. Given these results, we recommend applying an inverse method, which is discussed in Section 5.3.2.3.

Summary of Validation Diagnostic Tools for DCLS, Example 2: The spectra from four pure components were measured and were used with a simple baseline offset in a DCLS model. Five hundred and fifty measurement variables were used in the wavelength range of 1100–2198 nm. Eighteen samples with a range from 0 to 0.62 weight fraction for each component were used as a validation set. Statistical prediction errors were larger than expected from estimates made using spectral replicates, the predicted versus known plot revealed significant bias, and the spectral residuals showed significant structure. Given these validation results the model was deemed unreliable.

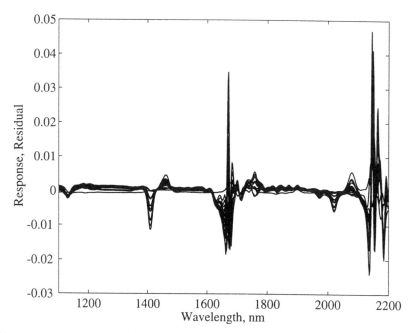

FIGURE 5.31. Spectral residuals from DCLS analysis of the four-component mixture.

Habit 5. Use the Model for Prediction

Habit 6. Validate the Prediction
Because the model is not acceptable, no predictions are performed.

5.2.2 Indirect CLS

In contrast to DCLS, the pure spectra in the indirect approach are not measured directly, but are estimated from mixture spectra. One reason for using
ICLS is that it is not possible to physically separate the components (e.g.,
when one of the components of interest is a gas and future prediction samples
are mixtures of the gas dissolved in a liquid). Indirect CLS is also used when
the model assumptions do not hold if the pure component is run neat. By
preparing mixtures, it is possible to dilute a strongly absorbing component so
that the model assumptions hold.

To estimate the pure spectra using the ICLS approach, a series of mixture
spectra are obtained based on an experimental design with known concentration values (C). The mixture spectra (R) are measured and related to the desired pure spectra (S) according to the equation $R = CS$. See the introduction
to Section 5.2 for development of this equation. Given the known R and C matrices, it is possible to estimate the pure spectra (S) using the following equation:

$$(C^TC)^{-1} C^TR = \hat{S} \qquad (5.16)$$

The reason to use designed experiments is to insure that the inverse of C^TC
can be computed. This means having a sufficient number of samples that are
mathematically independent. One requirement is that there must be at least as
many independent mixture samples as there are pure spectra to estimate. A
special class of designs called mixture designs must also be used (Cornell,
1990) in systems where solutions are being prepared to contain components
summing to a constant value ([e.g., 0.30 A, 0.20 B, and 0.50 C for a total of 1.0
(or 100%)].

See Section 5.2.1 for more discussion about CLS methods in general.

5.2.2.1 ICLS Example 1   The first ICLS example involves samples that contain three components—A, B, and C. The ranges of interest are 0.1–0.5 mol
fraction (A), 0.2–0.6 mol fraction (B), and 0.3–0.7 mol fraction (C). An extreme vertices mixture design is used to formulate the concentration matrix.
Table 5.7 and Figure 5.32 display the concentrations of the calibration samples
that are used to estimate the pure spectra. Spectra comprised of 200 measurement variables are obtained from these mixture samples. The pure spectra are
estimated from these mixtures following the "Six Habits of an Effective
Chemometrican" which are detailed in Chapter 1.

**TABLE 5.7. Three-Component Extreme Vertices Mixture Design.
Concentrations of Components A, B, and C**

Experiment No.	A	B	C
1	0.50	0.20	0.30
2	0.10	0.20	0.70
3	0.10	0.60	0.30
4	0.10	0.40	0.50
5	0.30	0.20	0.50
6	0.30	0.40	0.30
7	0.23	0.33	0.43
8	0.23	0.33	0.43
9	0.23	0.33	0.43

Habit 1. Examine the Data

The calibration spectra displayed in Figure 5.33 reveal a random baseline offset for these data. No other anomalous behavior is observed. The concentration design has seven independent standards (see Figure 5.32), which are sufficient for estimating the three pure spectra.

Habit 2. Preprocess as Needed

The varying baseline observed in Habit 1 should be removed by preprocessing. In the DCLS section (Section 5.2.1.1) the addition of a vector of ones to
the pure spectra matrix (**S**) was presented as a way to account for an offset

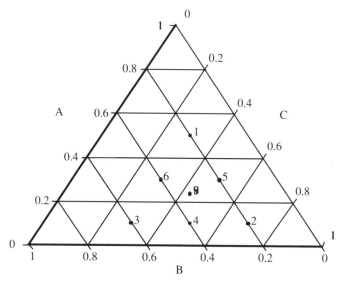

FIGURE 5.32. Ternary plot of the extreme vertices experimental design for ICLS
Example 1.

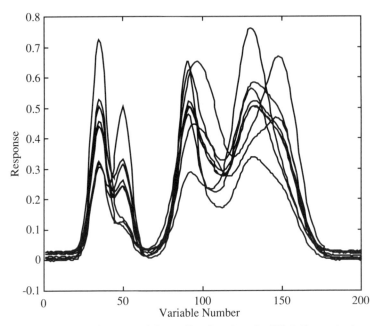

FIGURE 5.33. Spectra of the calibration data for ICLS Example 1.

when predicting unknown samples. With ICLS this approach cannot be used, because the initial goal in ICLS is to estimate the pure spectra (**S**). The baseline offset therefore must be removed by preprocessing (see Section 3.1.4). For these data, an average of the first 10 points of each spectrum is computed and subtracted from every value in the spectrum. If a baseline is also present in prediction samples, an appropriate baseline correction can then be added to the **S** matrix for prediction in the same manner as with DCLS.

Habit 3. Estimate the Model

To estimate the pure component spectra, the mixture spectra and concentrations of the calibration samples are supplied to the computer and a regression is performed as described by Equation 5.16.

Habit 4. Examine the Results/Validate the Model

There are two steps to validating an ICLS model. The first is to verify that the estimated pure spectra are reasonable; the CLS model assumptions are then validated. Several diagnostic tools for validating the pure spectra are discussed, and a summary is found at the end of the section in Table 5.8. The primary use of these diagnostic tools is to investigate whether the estimated pure spectra are reasonable.

Estimated Pure Spectra Plot (Model Diagnostic): Figure 5.34 shows the three estimated pure spectra for this example. These are evaluated given

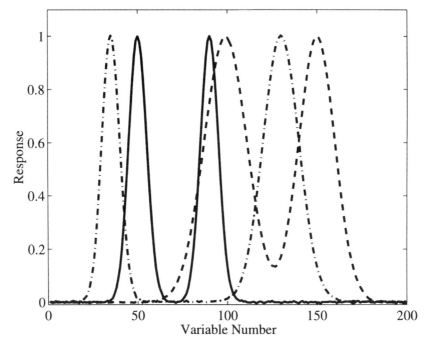

FIGURE 5.34. Estimated pure spectra using the baseline-corrected data. Solid, component A; dashed, component B; and dashed-dotted, component C.

knowledge about the chemistry of the samples and the measurement system. Based on the mixture spectra in Figure 5.33, the estimated pure spectra appear reasonable.

Uncertainty in Pure Spectra (Model Diagnostic): The pure-component spectra are estimated from a standard multiple linear regression calculation (Equation 5.16) and, therefore, error estimates are available. The error estimates for all pure spectra at variable λ are shown in Equation 5.17:

$$\mathrm{sd}(\mathbf{S}_\lambda) = \sqrt{\mathrm{diag}(\mathbf{C}^\mathrm{T}\mathbf{C})^{-1} s_\lambda^2} \tag{5.17}$$

where s_λ^2 is a measure of how well the estimated pure spectra reproduce the mixture spectra at variable λ.

The pure spectrum plot for component B with a ± 2 standard deviation uncertainty band is shown in Figure 5.35. The width of this band can be used to assess the quality of the estimated pure spectrum. The uncertainties are very small, indicating that the spectrum of pure component B is well determined by the calibration data. The same is observed with the other two components (not shown).

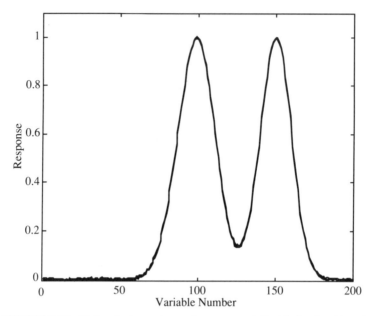

FIGURE 5.35. Estimated spectrum of component B with the 2-sd band.

Calibration Measurement Residual Plot (Model Diagnostic): After the pure spectra are estimated ($\hat{\mathbf{S}}$), they are used with the original $\mathbf{C}$ matrix to generate estimates of the mixture spectra ($\hat{\mathbf{R}} = \mathbf{C}\hat{\mathbf{S}}$). These are then used to calculate a calibration residual matrix which contains the portion of the mixture spectra that are not fit by the estimated pures (Equation 5.18).

$$\text{Calibration residual} = \mathbf{R} - \hat{\mathbf{R}} = \mathbf{R} - \mathbf{C}\hat{\mathbf{S}} \qquad (5.18)$$

The residuals shown in Figure 5.36 are randomly distributed about zero, which is expected when the model is correctly specified. The magnitude of the residuals is small relative to the size of the features found in the plot of mixture spectra and estimated pures.

In many applications, the spectral residuals will not behave in the ideal manner as depicted here. Some nonrandom behavior may be tolerable depending on the performance requirements of the model. If the residuals are large and the model performance is not acceptable, an inverse model approach such as PLS or PCR (Section 5.3) can be considered.

Summary of Validation Diagnostic Tools for ICLS, Example 1: Pure spectra for three components, A, B, and C, were estimated using mixtures. The details of the data set used to estimate the pure spectra are as follows:

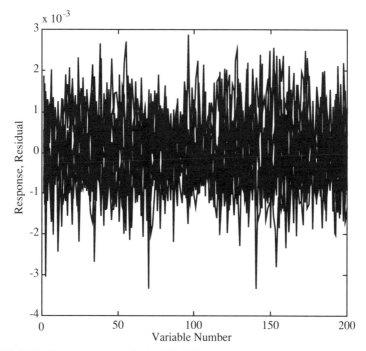

FIGURE 5.36. Spectral residuals for ICLS Example 1, accounting for baseline off-set.

Calibration Design: 9 samples, selected using a mixture design

Preprocessing: baseline correction using the average of the first 10 measurement variables.

Variable Range: 200 measurement variables

The operating ranges of the prediction model based on the calibration data follow. Predicting future samples from outside this operating range is extrapolating and may produce unreliable results.

A: 0.1–0.5

B: 0.2–0.6

C: 0.3–0.7

The conclusion is that the estimated pure spectra are reasonable based on the diagnostics summarized in Table 5.8. The first column in the table lists the name of each tool and the second column describes results from both well-behaved and problematic data.

TABLE 5.8. Summary of Validation Diagnostics for Estimating Pure Spectra Using the ICLS Method

Model Diagnostics	Description and Use
Estimated pure spectra plot ($\hat{S}$ vs. variable number)	Plot of the pure spectra estimated from the calibration samples.
	Used to validate the calculated pure spectra.
	The estimated pure spectra should have features that are expected given the measurement system used. For example, positive peaks are expected when using UV or IR spectroscopy.
Uncertainty in pure spectra [($\hat{S}$ ±2 sd band) vs. variable number]	The estimated pure spectra are plotted with a ±2 sd uncertainty band. The bands are a measure of the confidence in the pure spectra over all regions.
	The width of the band can be used to assess the quality of the estimated pure spectra.
	Small bands indicate that the estimated spectra are well determined by the calibration data.
	Good calibration experimental designs are necessary (but not sufficient) to minimize the uncertainty bands.
Calibration measurement residual plot [($\mathbf{r} - \hat{\mathbf{r}}$) vs. variable number]	The residuals are the portion of the calibration measurement vectors that are not fit by the estimated pure spectra and the known concentrations.
	Used to validate the estimated pure spectra.
	Residual values are expected to vary randomly about a line of zero slope and zero intercept.
	Patterns in the spectral residuals indicate an inadequacy in the model and/or relatively large calibration concentration errors.

Once the pure spectra have been estimated and validated, the CLS model is validated following the approach discussed for DCLS (Section 5.2.1.1). This can be done with the calibration set of mixtures or with an additional validation set. The validation of the pure spectra already gives some confidence in the model because violations of the CLS assumptions will be identified when reasonable pure spectra are not obtained. However, the DCLS validation tools found in Table 5.1 should still be used. For this example, this second step in the model validation did not reveal any problems with the model (results not shown). See Example 2 for discussion on the complete ICLS model validation.

Habit 5. Use the Model for Prediction
See Habit 5 for DCLS, Section 5.2.1.1.

Habit 6. Validate the Prediction
See Habit 6 for DCLS, Section 5.2.1.1.

"What If" 1: Baseline Offset

In this "what if," the pure component spectra for Example 1 are estimated without first removing the baseline. The following results are observed: (1) The resulting estimated pure spectra shown in Figure 5.37 have an offset. (2) The plot of the pure spectrum of component B with the ± 2 sd uncertainty band is shown in Figure 5.38. The 2-sd band is quite large and nearly constant across all variables. The width of the band relative to the features in the spectra qualitatively indicates a problem with the estimated pure. (3) The calibration residuals shown in Figure 5.39 look reasonably random except for the presence of a baseline offset.

This is in contrast to a small ± 2 sd uncertainty band (Figure 5.35) and random calibration residuals (Figure 5.36) when the baseline was removed.

"What If" 2: Transcription Error

In this "what if," the pure component spectra for Example 1 are estimated with a transcription error in the concentration matrix. Assume that for sample 3, component A was entered as 0.20 instead of 0.10 (see Table 5.7). Because the error is only for component A, it might be imagined that the error will only affect the estimation of the pure spectrum for this component. This is not correct, because the entire $\mathbf{C}$ matrix is used to estimate all of the pure spectra.

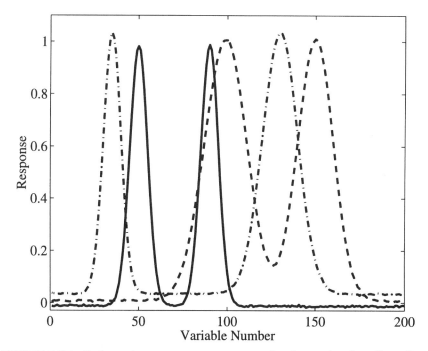

FIGURE 5.37. Estimated pure spectra, not accounting for random baseline offset. Solid, component A; dashed, component B; and dashed-dotted, component C.

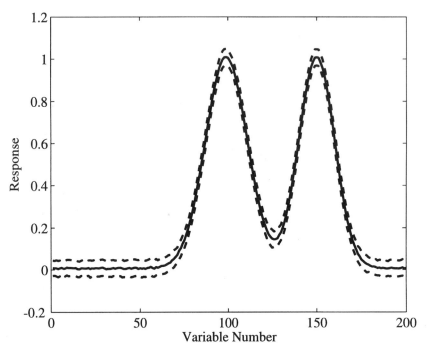

FIGURE 5.38. Estimated spectrum of component B (solid line) with the 2-sd deviation band (dashed lines), derived from data without baseline correction.

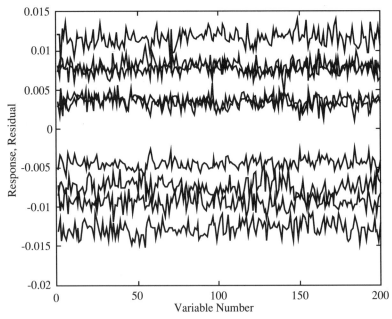

FIGURE 5.39. Spectral residuals for ICLS Example 1, not accounting for baseline offset.

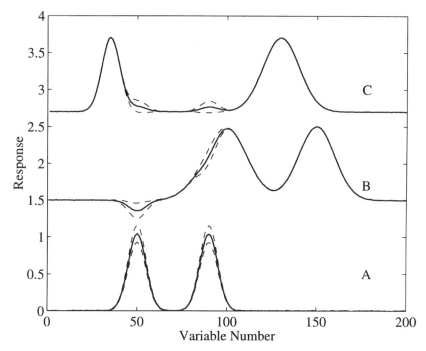

FIGURE 5.40. Estimated pure spectra with the 2-sd band for the case where there is a transcription error. Pure spectra are offset for clarity.

Figure 5.40 displays the pure spectra for components A, B, and C with the ±2 sd uncertainty bands. There are two regions in the pure spectra that have wide bands which coincide with the peaks in the pure spectrum of component A. The calibration spectral residuals shown in Figure 5.41 also clearly indicate a problem. With only these diagnostics, it is not possible to identify the problem. For diagnosing problems with samples and variables, see the DCLS examples and Table 5.1. (An additional simple step is to verify that the concentrations add up to 1.0 for each sample.)

5.2.2.2 ICLS Example 2 This example discusses the determination of sodium hydroxide (caustic) concentration in an aqueous sample containing sodium hydroxide and a salt using NIR spectroscopy. An example of this problem in a chemical process occurs in process scrubbers where CO_2 is converted to Na_2CO_3 and H_2S is converted to Na_2S in the presence of caustic. Although caustic and salts have no distinct bands in the NIR, it has been demonstrated that they perturb the shape of the water bands (Watson and Baughman, 1984; Phelan et al., 1989). Near-infrared spectroscopy is therefore a viable measurement technique. This method also has advantages as an analytical technique for process analysis because of the stability of the instrumentation and the ability to use fiber-optic probes to multiplex the interferometers and locate them remotely from the processes.

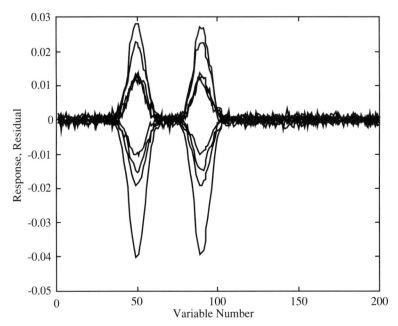

FIGURE 5.41. Spectral residuals from the case in which there is a transcription error.

The goal of the project is to determine the concentration of caustic in an aqueous stream containing 6–11% caustic and 12–17% salt. The requirement for the caustic determination is to predict the concentration to ±0.1 wt.% (1σ). Because NIR spectroscopy of aqueous systems is known to be sensitive to temperature fluctuations, another factor that must be considered is the temperature of the sample, which can vary from 50 to 70°C.

This is an unusual application of CLS modeling because caustic and salt do not have distinct spectral features in the NIR. It is interesting to see if the perturbations in the water band behave linearly with respect to concentration. Twelve calibration standards were prepared in the laboratory according to a constrained-mixture experimental design (Figure 5.42). These standards were slowly heated from 50 to 70°C while collecting spectra (using a 1-cm pathlength). This resulted in 6–10 spectra for each calibration standard, for a total of 95 spectra. (These data are also discussed in Sections 5.3.1.2 and 5.3.2.2.)

Habit 1. Examine the Data

The experimental design of the 12 standards is plotted in Figure 5.42. They cover the range of interest and appear to be well distributed within the space. On this graph the concentrations are presented in weight fractions, but the calculations are performed using weight percent. The data have been split into calibration and validation sets. The calibration samples (indicated by the X) are

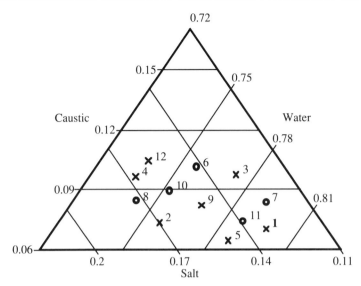

FIGURE 5.42. Experimental design for ICLS Example 2: × = calibration; ○ = validation.

used to estimate and validate the pure spectra, and the validation samples (indicated by ○) are used to validate the CLS model.

The 95 calibration and validation spectra collected from the 12 samples at varying temperatures are shown in Figure 5.43a. The 7500 to 4000 cm^{-1} region is not useful because of the extremely high absorbance values. Furthermore, the spectral region above 9500 cm^{-1} is excluded owing to poor sensitivity of the detector. No unusual samples or variables are noted in the expansion of the 9500 to 7500-cm^{-1} region (Figure 5.43b).

Habit 2. Preprocess as Needed
Because there is no peak located at ~9500 cm^{-1}, the variation in this region indicates a fluctuating baseline (see Figure 5.43b). This can be removed by correction at a single wavelength or by using a first derivative (see Section 3.1.4). Both approaches were considered and the first derivative gave superior results. The 9087 to 8315-cm^{-1} region consisting of 101 responses was selected after further elimination of wavelengths with low variability. The resulting preprocessed data are shown in Figure 5.44. Note that taking the derivative changes the y-axis scale (this is especially important to remember when examining the magnitude of spectral residuals).

Habit 3. Estimate the Model
To estimate the pure-component spectra for caustic, salt, and water, the preprocessed spectra and concentrations of the calibration mixture samples are supplied to the computer.

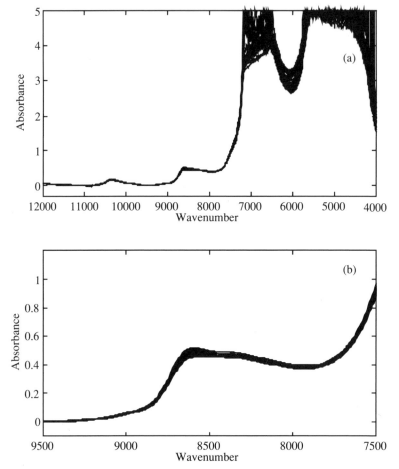

FIGURE 5.43. Raw data from (a) 12000–4000 cm^{-1} and (b) 9500–7500 cm^{-1}.

Habit 4. Examine the Results/Validate the Model

Estimated Pure Spectra Plot (Model Diagnostic): The estimated pure spectra shown in Figure 5.45 reveal a caustic pure spectrum that has a negative peak. With NIR the pure spectra are expected to have positive bands. However, in this example, this expectation is not reasonable because peak perturbations are being modeled and a first derivative has been used.

Uncertainty in Pure Spectra (Model Diagnostic): The caustic spectrum and uncertainties plotted in Figure 5.46 reveal large uncertainties throughout the spectral region. The uncertainties in the other pure-component spectra show similar results.

Calibration Measurement Residual Plot (Model Diagnostic): The magnitude of the calibration spectral residuals shown in Figure 5.47 are large relative to the original preprocessed data (Figure 5.44), and they also have nonrandom features. These observations indicate a potential problem with the

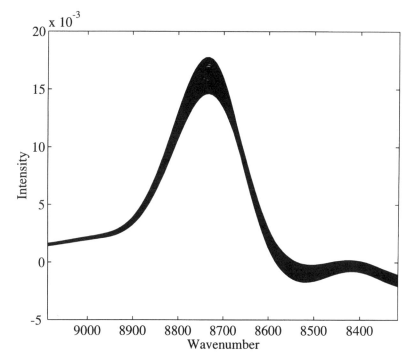

FIGURE 5.44. First derivative data plotted from 9087 to 8316 cm^{-1}.

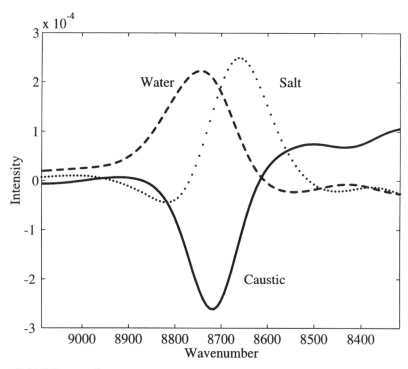

FIGURE 5.45. Estimated pure spectra without accounting for temperature.

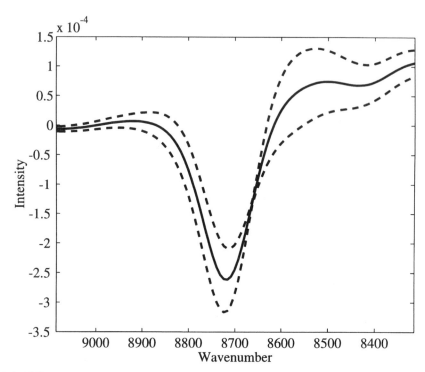

FIGURE 5.46. Estimated pure spectrum of caustic (solid line) with the 2-sd bands (dashed lines).

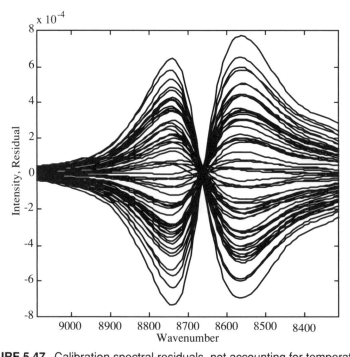

FIGURE 5.47. Calibration spectral residuals, not accounting for temperature.

model. It is possible that the perturbations in the water band do not vary linearly with concentration, or that the effect of temperature on the spectra (which has not been included in the model) is significant.

There appear to be problems with this ICLS model that need to be identified and corrected. A few of the diagnostics from the model validation (Table 5.1) are used to investigate the problem. Figure 5.48 shows the predicted versus known caustic concentration for the validation samples using the pure spectra derived from the calibration data. The concentration predictions are poor. There is no apparent structure in the errors with respect to caustic concentration, which implies that the problem is not a nonlinear relationship between the spectra and concentration. Because temperature has not been included in the model, it is postulated that the errors are related to this parameter. Figure 5.49 is a plot of the caustic concentration residual versus temperature and reveals a systematic change in the concentration residuals with temperature. This indicates that temperature has a significant effect on the spectra that is not being accounted for by the model. To confirm this, the preprocessed spectra from one sample over all temperatures are plotted in Figure 5.50. Comparing the spread of the spectra in this figure with that found in Figure 5.44 indicates a large variance contribution from temperature. The conclusion is that the temperature effect needs to be incorporated into the model.

Adding a physical term such as temperature to the model is unusual for CLS modeling. It will only be successful if the CLS assumptions of the model hold for all pure components (including temperature). Temperature is included in

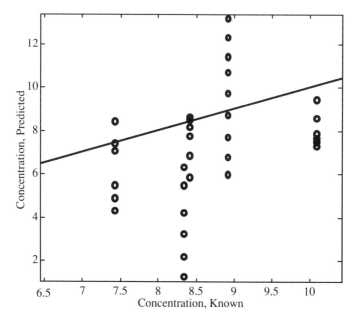

FIGURE 5.48. Predicted versus actual concentrations for the validation data, not accounting for the temperature.

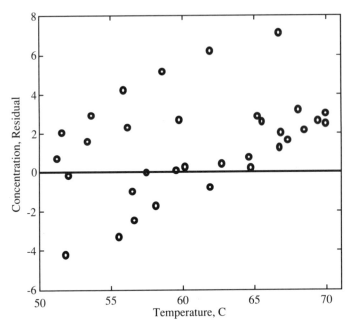

FIGURE 5.49. Concentration residuals of the validation samples as a function of temperature.

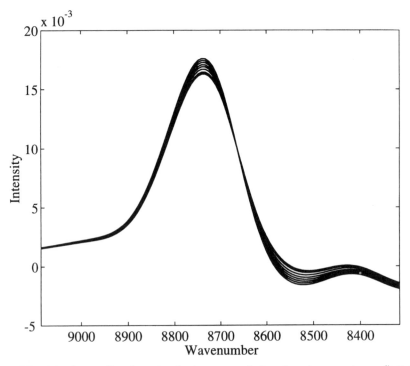

FIGURE 5.50. One calibration standard measured at various temperatures, first derivative.

the model by adding another column to the **C** matrix containing the sample temperature during data collection.

Estimated Pure Spectra Plot (Model Diagnostic): The estimated pure spectra for this expanded model are shown in Figure 5.51. Four pure spectra are shown (three chemical components plus temperature).

Uncertainty in Pure Spectra (Model Diagnostic): The caustic pure spectrum uncertainties shown in Figure 5.52 are smaller than with the previous model (Figure 5.46).

Calibration Measurement Residuals Plot (Model Diagnostic): The calibration spectral residuals shown in Figure 5.53 are still structured, but are a factor of 4 smaller than the residuals when temperature was not part of the model. Comparing with Figure 5.51, the residuals structure resembles the estimated pure spectrum of temperature. Recall that the calibration spectral residuals are a function of model error as well as errors in the concentration matrix (see Equation 5.18). Either of these errors can cause nonrandom features in the spectral residuals. The temperature measurement is less precise relative to the chemical concentrations and, therefore, the hypothesis is that the structure in the residuals is due to temperature errors rather than an inadequacy in the model.

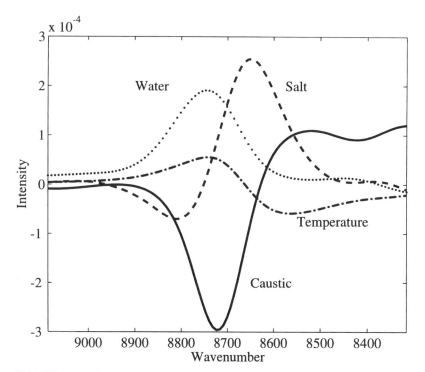

FIGURE 5.51. Pure spectra estimated when accounting for the temperature.

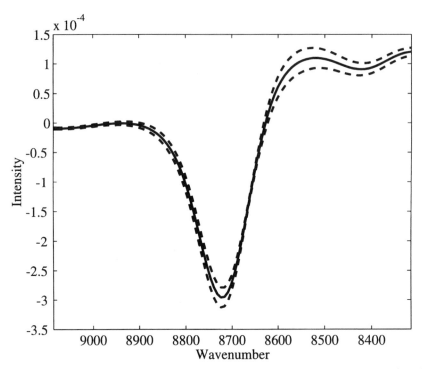

FIGURE 5.52. Estimated pure spectrum of caustic (solid line) with the 2-sd bands (dashed lines) when accounting for temperature.

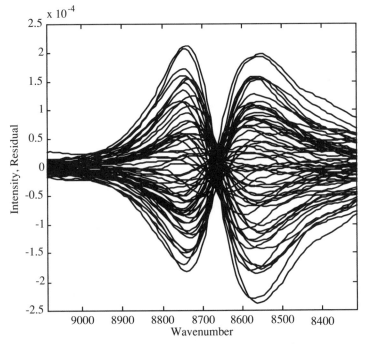

FIGURE 5.53. Calibration spectral residuals from estimating the pure spectra, accounting for the temperature.

The estimated pure spectra are accepted and the ICLS model is validated using the diagnostics in Table 5.1, Section 5.2.1.1. The validation data are obtained by performing prediction on the validation samples using the pure spectra estimated from the calibration samples.

Statistical Prediction Errors (Model and Sample Diagnostic): From the **S** matrix it is possible to predict all four components (caustic, salt, water concentration, and temperature). However, in this application the interest is only in caustic and, therefore, only the results for this component are presented. The statistical prediction errors for the caustic concentration for the validation data vary from 0.006 to 0.028 wt.% (see Figure 5.54). The goal is to predict the caustic concentration to 0.1 wt.% (1σ), and the statistical prediction errors indicate that the precision of the method is adequate. Also, there do not appear to be any sample(s) that have an unusual error when compared to the rest of the samples.

Predicted vs. Known Concentration Plot (Model and Sample Diagnostic): The predicted versus known concentrations plotted in Figure 5.55 are greatly improved compared to the initial model (Figure 5.48). The points are close to the ideal line, indicating no problems with the model. Furthermore, there do not appear to be any sample(s) that are unusually far from the line than the rest of the data.

Concentration Residuals vs. Predicted Concentration Plot (Model and Sample Diagnostic): The concentration residuals are plotted as a function of the predicted caustic concentration in Figure 5.56. The concentration errors

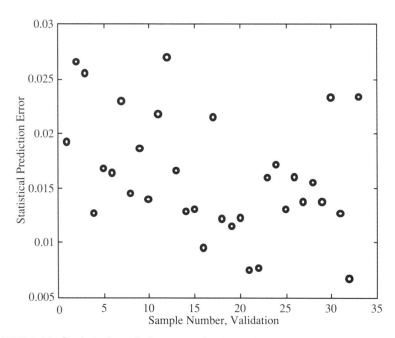

FIGURE 5.54. Statistical prediction errors for the validation samples (wt.% caustic), accounting for the temperature.

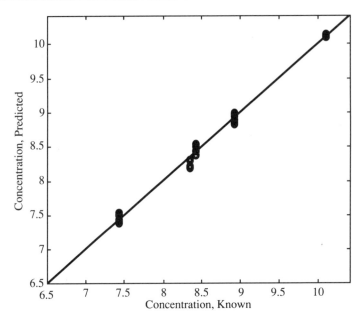

FIGURE 5.55. Predicted versus known concentrations for the validation samples (caustic), after taking the temperature into account.

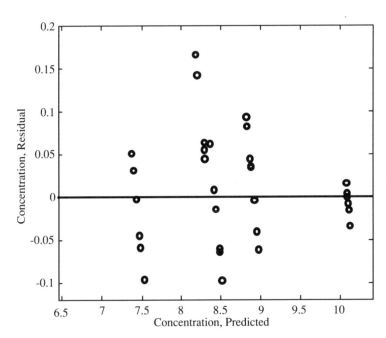

FIGURE 5.56. Concentration residuals for the validation samples versus predicted concentration (caustic), accounting for the temperature.

range from -0.10 to 0.17 wt.%, which is within the performance required by the application. The residuals do not appear to be structured as a function of the caustic concentration, indicating that the model has adequately captured the caustic variation.

Concentration Residuals vs. Other Postulated Parameter Plot (Model and Sample Diagnostic): The caustic concentration residuals are plotted as a function of temperature in Figure 5.57 to verify that the temperature has been modeled. There does not appear to be any structure in this plot, indicating that temperature has been adequately incorporated into the model.

Root Mean Square Error of Prediction (RMSEP) (Model Diagnostic): The RMSEP for the determination of caustic is 0.06 wt.% over a range of 7.4–10.4 wt.%. This estimate of prediction ability indicates that the performance of the model is acceptable for the application.

Measurement Residual Plot (Model, Sample and Variable Diagnostic): The spectral residuals for the validation data shown in Figure 5.58 are an order of magnitude smaller and less structured than the residuals obtained when the pure spectra were estimated (Figure 5.53). This can be explained as follows: Equation 5.18 shows that the reported concentrations and temperatures (C is used in Equation 5.18) are used in the computation of the calibration residuals. Therefore, errors in the reported concentrations and temperatures contribute to the calibration residuals in addition to model inadequacy. In contrast, estimated concentrations and temperatures ($\hat{C}$ is used in Equation 5.13)

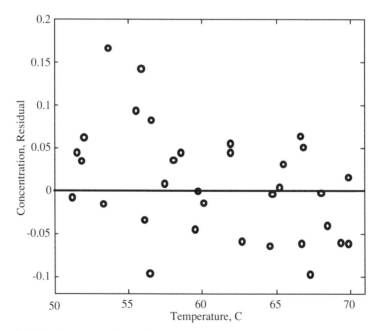

FIGURE 5.57. Concentration residuals for validation samples as a function of the sample temperature (caustic), accounting for the temperature.

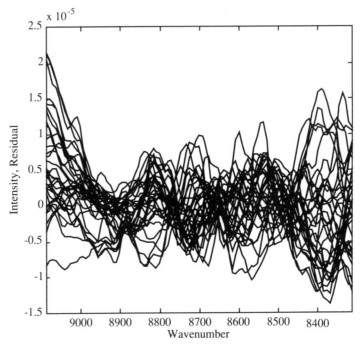

FIGURE 5.58. Spectral residuals for the validation data, accounting for temperature.

are used for the validation residuals. The estimated temperatures are derived from the spectral measurements, which provide a more precise estimate of the temperature than the reference method (thermocouple). The error in the temperature is therefore reduced, which is reflected in a reduction in the magnitude of the spectral residuals. The random nature of these residuals corroborates the earlier hypothesis that the model is correct and the structure in the calibration residuals is the result of errors in the reported temperature.

Summary of Validation Diagnostic Tools for ICLS, Example 2: The determination of caustic concentration in an aqueous sample containing NaOH (caustic) and salt using NIR spectroscopy is examined in this example. The goal was to create a model to predict caustic concentration in aqueous samples with varying salt and temperature levels. The details of the data set used to construct the calibration model are as follows:

Calibration design: 62 spectra, 7 design points (collected with varying temperature)

Validation design: 33 spectra, 5 design points (collected with varying temperature)

Preprocessing: 15-point first derivative

Variable range: 101 original measurement variables, from 9087 to 8315 cm^{-1}

The operating ranges of the prediction model based on the calibration data follow. Predicting future samples from outside this operating range is extrapolating and may produce unreliable results.

Caustic: 6.4–10.4 wt.%

Salt: 13.0–17.0 wt.%

Temperature: 50.2–69.8°C

The ICLS model contains pure spectra of caustic, salt, water, and temperature. The conclusion of the model validation is that the ICLS model adequately describes the water peak shape changes due to caustic, salt, and temperature. Furthermore, it meets the required performance criterion for the prediction of caustic. The measures of performance are as follows (see Table 5.1 for a description of these figures of merit):

RMSEP = 0.06 wt.%

Maximum statistical prediction error = 0.028 wt.%

Expected error in prediction = ±0.17 wt.%

(Based on errors in predicted caustic concentration. See Figure 5.56.)

Habit 5. Use the Model for Prediction

The predicted caustic concentrations for 99 prediction samples are plotted in Figure 5.59.

Habit 6. Validate the Prediction

The four prediction diagnostic tools found in Table 5.4 in Section 5.2.1.1 are applied to the prediction data. These tools are used to assess the reliability of the prediction results.

Predicted Concentration Plot: Figure 5.59 shows that all predicted caustic concentrations are within the range of the calibration data (shown as horizontal lines).

Statistical Prediction Errors: The statistical prediction errors are plotted in Figure 5.60. The maximum from the model validation is indicated by a horizontal line. There are a few samples above this maximum and one sample (54) that has an error considerably larger than the rest. The measurement residuals for these samples will be investigated further.

Measurement Residuals Plot: The spectral residuals for all prediction samples are plotted in Figure 5.61, with the validation residual range indicated by the horizontal lines. There are a few samples with residual features that extend slightly outside of the calibration range and one sample (54) with residual features that are clearly outside the calibration range. If this were a laboratory measurement, the analysis of sample 54 would be repeated. For an on-line measurement, the results from this sample would be considered suspect and would alert the operator to possible instrumental or process changes. The final validation diagnostic from Table 5.4 (raw measurement data plot) is not presented.

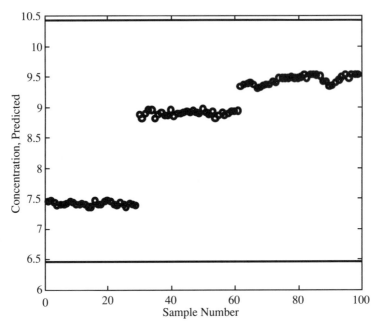

FIGURE 5.59. Predicted caustic concentrations for unknowns with the calibration ranges denoted by horizontal lines.

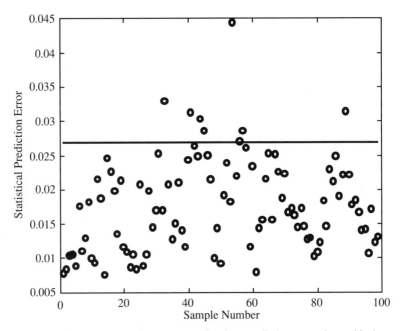

FIGURE 5.60. Statistical prediction errors for the prediction samples, with the maximum from validation indicated by the horizontal line.

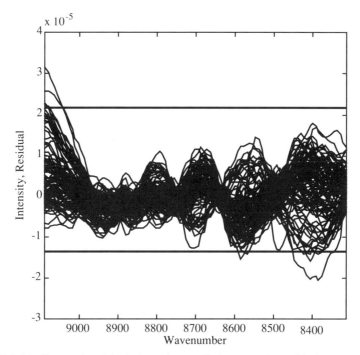

FIGURE 5.61. Spectral residuals from the prediction samples, with the range of calibration residuals shown by horizontal lines.

Summary of Prediction Diagnostic Tools for ICLS, Example 2: Based on the prediction diagnostics, the conclusion is that the predicted values for 98 of 99 prediction samples are reasonable. Based on the range of validation concentration residuals (see Figure 5.56), the errors in the predicted caustic concentrations of the unknowns are expected to be within ±0.17 wt.% corresponding to an RMSEP of 0.06 wt.%.

5.2.3 Summary of Classical Methods

Summary of DCLS

The DCLS method can be applied to simple systems where all of the pure-component spectra can be measured. To construct the DCLS model, the pure-component spectra are measured at unit concentration for each of the analytes in the mixture. These are used to form a matrix of pure spectra (**S**) and the model is then constructed as the pseudo-inverse of this **S** matrix. This calibration model is used to predict the concentrations in unknown samples.

Weaknesses

1. Strict assumptions must be obeyed for the method to perform well (i.e., measurements are linear with concentration, linear additivity of pure component spectra holds, all components are known).

2. Must be able to obtain the spectra of all of the pures components of interest.

3. Must explicitly include in the **S** matrix all of the sources of variation affecting the response.

Strengths

1. Prediction diagnostics are available that can be used to assess the reliability of predicted values. It is also possible to use the prediction diagnostics to postulate the cause of poor predictions.

2. Only a small number of samples are needed to construct the calibration model (i.e., only the pure spectra are measured).

3. It is a simple method to describe.

4. Because many redundant variables are used, it is possible to quantitate analytes with overlapping features.

5. It is a well-understood method with a solid statistical foundation.

Summary of ICLS

As with DCLS, ICLS uses pure spectra to develop a model for predicting concentration values from unknown spectra. However, with the indirect approach the pure spectra are not measured directly, but are estimated from mixture spectra. Once the pure spectra are estimated, ICLS uses the same steps as described for the DCLS approach to validate the resulting model.

Weaknesses

1. Experimental design for estimation of the pures is an issue when using ICLS.

2. More work is required to obtain the model as compared to DCLS.

3. See also weaknesses 1 and 3 of DCLS.

Strengths

1. Can estimate pure spectra when it is not possible to measure them directly.

2. The number of samples that need to be measured in the calibration phase is small as compared to inverse modeling (Section 5.3). The minimum number of mixtures to measure is equal to the number of analytes in the system.

3. In some situations, more effective estimates of the pure spectra are obtained than measuring the pure components directly (e.g., estimating pures within limited concentration ranges to insure linear response).

4. Additional validation diagnostics are generated during the estimation of the pure component spectra.

5. See also advantages 1 and 3–5 of DCLS.

Both DCLS and ICLS are methods that work well on simple systems. We only consider using these methods if the theory describing the analytical tool employed supports the model assumptions. The only difference between these two approaches is in how the pure spectra are obtained. Therefore, the choice of direct over indirect CLS is dictated by the availability of pure-component spectra.

5.3 INVERSE METHODS

In this section three inverse least squares (ILS) methods are discussed: multiple linear regression (MLR), partial least squares (PLS), and principal components regression (PCR). There are several reasons for choosing the path of ILS from the calibration decision tree (Figure 5.3). For example, CLS may not be appropriate because the system under investigation is not simple or it may not be possible to obtain the pure spectra of all the analytes expected to be in the unknown samples. In practice, it is not always clear as to which approach, classical versus inverse, is optimal. Researchers have studied these different approaches using simulated data and designed experiments and give some guidance on method selection (Haaland and Thomas, 1988).

The common feature for all of the inverse methods is in how the relationship between the measurements and concentration is modeled. The concentrations are treated as a function of the responses, as shown in the following equation:

$$\mathbf{c} = \mathbf{Rb} \tag{5.19}$$

where the (nsamp $\times$ 1) vector $\mathbf{c}$ contains the concentrations, $\mathbf{R}$ (nsamp $\times$ nvars) is a matrix of measurements, and the (nvars $\times$ 1) vector $\mathbf{b}$ contains the model coefficients. This equation can be used to model the relationship between multiple analytes of interest (different $\mathbf{c}$ vectors) and the same response matrix ($\mathbf{R}$) using different model coefficients ($\mathbf{b}$ vectors).

Comparing the inverse model found in Equation 5.19 to the model for the classical method (Equation 5.6, $\mathbf{r} = \mathbf{c} \ \mathbf{S}$), it may not be obvious that the approaches are significantly different. To illustrate the difference, the details of the matrix algebra for Equation 5.19 are presented for one sample:

$$[c_1] = [r_{11} \quad r_{12} \quad r_{13} \quad r_{14}] \begin{bmatrix} b_1 \\ b_2 \\ b_3 \\ b_4 \end{bmatrix} \tag{5.20}$$

Equation 5.20 shows that the concentration value (c_1) is equal to a weighted sum of the responses for different variables (i.e., $r_{11}{}^*b_1 + r_{12}{}^*b_2 + r_{13}{}^*b_3 + r_{14}{}^*b_4$) where the regression coefficients ($\mathbf{b}$) in Equation 5.19 are the weights.

This is different from the classical approach that fits a linear combination of the pure spectra to an observed unknown spectrum (see Equations 5.7–5.10).

This difference between the modeling approaches leads to some advantages for the inverse methods. For example, with the classical approaches it is necessary to model all of the components in the system simultaneously. This is illustrated in Section 5.2.2.2, where CLS fails to produce good predictions when all of the components in the system are not explicitly considered. With the ILS methods, it is possible to predict the concentration of one component even if additional chemical and physical sources of variation are present (e.g., interferents). The requirements are that the measurement system adequately differentiate the component of interest from other sources of variance, the instrument response is sufficiently linear with concentration, and the calibration experiments are carefully planned. This ability of the inverse methods to implicitly model all of the other sources of variation is very powerful. Practically, this means that much less explicit knowledge of the system is required when using the inverse methods, but more work may be required for calibration.

It is an oversimplification to say that one can be completely ignorant of the additional sources of variation in a system and still produce good models. In fact, it is very important to insure that all the significant sources of variation are present when the calibration models are estimated (i.e., good experimental design is required). However, these other sources of variance are not included as additional variables in the model (explicitly), but are implicitly modeled. Any source of variation that does not change during the calibration phase will not implicitly be included in the model. Therefore, the ideal approach is to manipulate all of the sources of variance using an experimental design to insure that they are all represented. If designed experiments are not feasible, another approach is to measure many samples and assume that the variance that is relevant to the prediction of future samples is present. This very important concept of experimental design for inverse modeling is discussed more fully in Appendix A of Chapter 2.

A simple example using both CLS and ILS is presented to illustrate the effect of an interferent on the models. Assume that it is believed that there are two pure components in a system of interest and a three-variable pure spectrum of each component has been measured ($s_1 = [3 \quad 2 \quad 1]$ and $s_2 = [1 \quad 2 \quad 3]$). In reality, there is an additional unknown interferent in the mixtures with the pure spectrum, $i_1 = [3 \quad 0 \quad 0]$.

Assume also that a validation sample has been collected with concentrations for s_1, s_2, and i_1 of 1, 3, and 2 respectively ($c = [1 \quad 3 \quad 2]$). Assuming linear additivity holds, the resulting response vector for this mixture sample is $r = [12 \quad 8 \quad 10]$. When validating the models using this sample, the known information is the measured spectrum, $r = [12 \quad 8 \quad 10]$, and the component concentrations for the known analytes $c = [1 \quad 3]$. The steps for validating the CLS model are shown in Figure 5.62 and include (a) formulating

a)	r	=	c	*	S
	$\begin{bmatrix} 12 & 8 & 10 \end{bmatrix}$	=	$[??]$	*	$\begin{bmatrix} 3 & 2 & 1 \\ 1 & 2 & 3 \end{bmatrix}$

b)	r	*	$S^{\dagger}$	=	$\hat{c}$
	$\begin{bmatrix} 12 & 8 & 10 \end{bmatrix}$	*	$\begin{bmatrix} 0.33 & -0.17 \\ 0.08 & 0.08 \\ -0.17 & 0.33 \end{bmatrix}$ =		$\begin{bmatrix} 3 & 2 \end{bmatrix}$

c)	$\hat{r}$	=	$\hat{c}$	*	S
	$\begin{bmatrix} 11 & 10 & 9 \end{bmatrix}$	=	$\begin{bmatrix} 3 & 2 \end{bmatrix}$	*	$\begin{bmatrix} 3 & 2 & 1 \\ 1 & 2 & 3 \end{bmatrix}$

d)	Spectral Residual	=	r	-	$\hat{r}$
	$\begin{bmatrix} 1 & -2 & 1 \end{bmatrix}$	=	$\begin{bmatrix} 12 & 8 & 10 \end{bmatrix}$	-	$\begin{bmatrix} 11 & 10 & 9 \end{bmatrix}$

e)	Conc. Residual	=	c	-	$\hat{c}$
	$\begin{bmatrix} 2 & 1 \end{bmatrix}$	=	$\begin{bmatrix} 1 & 3 \end{bmatrix}$	-	$\begin{bmatrix} 3 & 2 \end{bmatrix}$

FIGURE 5.62. Example of calibration and validation using the classical calibration approach. *(a)* Initial classical model form; *(b)* estimating concentrations; *(c)* reconstructing the response vector; *(d)* calculating the spectral residual; *(e)* calculating the concentrational residual.

the model, (b) estimating the concentrations in the validation sample, (c) estimating the unknown response vector, (d) calculating the spectral residuals, and (e) calculating the concentration residuals.

The nonzero spectral residuals in step (d) indicate a problem (no noise was added to the data and therefore zero residuals are expected). The concentration residuals are also larger than the expected errors (again zero noise), further indicating a problem. In summary, the classical approach has failed to produce the correct concentration values, but the diagnostic tools flagged a problem.

To apply the inverse approach, a set of calibration samples must be collected using an appropriate experimental design. For this example, assume the following concentration matrix represents a reasonable design.

$$
\mathbf{C} = \begin{array}{ccc} \mathbf{c}_1 & \mathbf{c}_2 & \text{int} \\ \begin{bmatrix} 1.0 & 2.0 & 0.0 \\ 1.2 & 4.0 & 0.0 \\ 0.1 & 0.4 & 2.0 \end{bmatrix} \end{array} \qquad (5.21)
$$

As with the classical example, all calculations are performed without knowledge about the presence of the interferent ("int" in Equation 5.21). It is only by chance that the interferent is represented in the calibration design. In practice, many samples would be collected in order to increase the likelihood that the interfering species would be adequately represented in the design.

The concentration matrix given above yields the following response matrix:

$$
\mathbf{R} = \begin{array}{c} \\ \\ \text{nsamp} \end{array} \begin{bmatrix} 5.0 & 6.0 & 7.0 \\ 7.6 & 10.4 & 13.2 \\ 6.7 & 1 & 1.3 \end{bmatrix}^{\text{nvars}} \qquad (5.22)
$$

Using this response matrix and the known concentration of only one of the components (c), the regression coefficients in Equation 5.19 can be estimated as

$$
\hat{\mathbf{b}} = (\mathbf{R}^T\mathbf{R})^{-1}\mathbf{R}^T\mathbf{c} \qquad (5.23)
$$

where $\hat{\mathbf{b}}$ contains the estimated regression coefficients. In the literature $(\mathbf{R}^T\mathbf{R})^{-1}\mathbf{R}^T$ is sometimes abbreviated as $\mathbf{R}^{\dagger}$ (called the pseudo-inverse of $\mathbf{R}$). Equation 5.23 is then written as

$$
\hat{\mathbf{b}} = \mathbf{R}^{\dagger}\mathbf{c} \qquad (5.24)
$$

Given the spectrum of a single unknown sample $(\mathbf{r}_{unk})$ it is possible to use the estimated regression vector $(\hat{\mathbf{b}})$ to predict the concentration of the one component of interest as follows:

$$
\hat{c} = \mathbf{r}_{unk}\hat{\mathbf{b}} \qquad (5.25)
$$

Figure 5.63 displays the steps in calculating the model using the calibration matrix in Equation 5.22 and performing prediction on the same validation sample studied using the classical approach. Using the initial formulation of the inverse model shown in (a), the steps are as follows: (b) estimate the regression coefficients for the components (only the estimate of the regression vector for

a) $\mathbf{c}_1$ $=$ $\mathbf{R}$ $*$ $\mathbf{b}_1$

$$\begin{bmatrix} 1 \\ 1.2 \\ 0.1 \end{bmatrix} = \begin{bmatrix} 5.0 & 6.0 & 7.0 \\ 7.6 & 10.4 & 13.2 \\ 6.7 & 1 & 1.3 \end{bmatrix} * \begin{bmatrix} ? \\ ? \\ ? \end{bmatrix}$$

b) $\mathbf{R}^{\dagger}$ $*$ $\mathbf{c}_1$ $=$ $\hat{\mathbf{b}}_1$

$$\begin{bmatrix} 0.0083 & -0.0208 & 0.1667 \\ 2.0458 & -1.0521 & -0.3333 \\ -1.6167 & 0.9167 & 0.1667 \end{bmatrix} * \begin{bmatrix} 1 \\ 1.2 \\ 0.1 \end{bmatrix} = \begin{bmatrix} 0.0000 \\ 0.7500 \\ -0.5000 \end{bmatrix}$$

c)

Component 1	Component 2

$\mathbf{c}_1 = \mathbf{r} * \hat{\mathbf{b}}_1$ | $\mathbf{c}_2 = \mathbf{r} * \hat{\mathbf{b}}_2$

$$[1] = \begin{bmatrix} 12 & 8 & 10 \end{bmatrix} * \begin{bmatrix} 0.0000 \\ 0.7500 \\ -0.5000 \end{bmatrix}$$
$$[3] = \begin{bmatrix} 12 & 8 & 10 \end{bmatrix} * \begin{bmatrix} 0.0000 \\ -0.2500 \\ 0.5000 \end{bmatrix}$$

FIGURE 5.63. Example of calibration and validation using the inverse calibration approach. *(a)* Initial inverse model form; *(b)* estimating the regression vector; *(c)* predicting the concentrations of components 1 and 2.

component 1 is shown), and (c) estimate the concentrations of the two components using the two estimated regression vectors.

Recall from the earlier discussion, the validation spectrum $\mathbf{r} = [12 \quad 8 \quad 10]$ is the result of a concentration vector of $\mathbf{c} = [1 \quad 3 \quad 2]$ for components 1, 2, and the interferent, respectively. Figure 5.63c shows that the inverse method has done a perfect job of predicting the concentration of the two components in the validation sample without using the interferent concentration in any of the calculations.

How can the inverse method correct for the interferent when it was not explicitly included in the model? For this example, it is easy to see. Recall that the spectrum of the interferent is $\mathbf{i}_1 = [3 \quad 0 \quad 0]$. The estimated regression vectors ($\hat{\mathbf{b}}$) in Figure 5.63c have zeros for the variable on which the interferent responds (variable 1). In this case, the inverse approach has implicitly modeled the presence of the interferent by ignoring the response variable that is associated with the interfering component. This example demonstrates that, for this well-behaved linear example, the inverse method accounts for the interferent provided it is present in at least one of the calibration samples. In more complex systems, the interferent must be present at varying levels in more than one sample before its effect can be adequately modeled. In general, the inverse methods will not selectively eliminate response variables where the interferences have signal (Seasholtz and Kowalski, 1990). This example was presented only to demonstrate that the inverse methods can compensate for the interferent.

With this background information on the inverse methods, it is instructive to examine the calculations for the inverse model in more detail. In Equation 5.23, the key to the model-building step is the inversion of the matrix $(\mathbf{R}^T\mathbf{R})$. This is a square matrix with number of rows and columns equal to the number of measurement variables (nvars). From theory, a number of independent samples in the calibration set greater than or equal to nvars is needed in order to invert this matrix. For most analytical measurement systems, nvars (e.g., number of wavelengths) is greater than the number of independent samples and therefore $\mathbf{R}^T\mathbf{R}$ cannot be directly inverted. However, with a transformation, calculating the pseudo-inverse of $\mathbf{R}$ $(\mathbf{R}^\dagger)$ is possible. How this transformation is accomplished distinguishes the different inverse methods.

Multiple linear regression with variable selection makes the matrix inversion possible by selecting a subset of the original variables. Both PCR and PLS reduce the number of variables by calculating linear combinations of the original variables (factors) and using a small enough number of these factors to allow for the matrix inversion.

5.3.1 Multiple Linear Regression (MLR)

As discussed in the introduction, the solution of the inverse model equation for the regression vector involves the inversion of $\mathbf{R}^T\mathbf{R}$ (see Equation 5.23). In many analytical chemistry experiments, a large number of variables are measured and $\mathbf{R}^T\mathbf{R}$ cannot be inverted (i.e., it is singular). One approach to solving this problem is called stepwise MLR where a subset of variables is selected such that $\mathbf{R}^T\mathbf{R}$ is not singular. There must be at least as many variables selected as there are chemical components in the system and these variables must represent different sources of variation. Additional variables are required if there are other sources of variation (chemical or physical) that need to be modeled. It may also be the case that a sufficiently small number of variables are measured so that MLR can be used without variable selection.

The only completely rigorous method to determine an optimum subset of variables is to test all possible combinations. The problem with this approach is that even with as few as 100 variables the number of combinations becomes too large to consider. For example, to select an optimum subset of three variables from amongst 100 measurement variables, 161,700 different combinations need to be tested. Because of the large number of possibilities, various strategies have been developed to select variables with a reasonable amount of work. The most common approaches include: step forward, step backward, and mixed (step forward with a backward glance). The discussion about MLR and stepwise approaches found in this chapter serves as an introduction to this area of statistics. The interested reader is referred to the literature for more rigorous discussions of this topic (Weisberg, 1985).

The approach for variable selection used in this chapter is the mixed method. It begins with a model containing only the intercept term. Variables that most improve the fit to the data are then added to the model one at a time, provided that they have statistically significant regression terms. The method also removes a variable if it becomes statistically insignificant when subsequent variables are added. The levels of significance for adding and re-moving variables are specified by selecting the "Probability to Enter" and the "Probability to Leave."

The advantage of estimating a model with stepwise MLR rather than with the "full-spectrum" techniques (e.g., PLS and PCR, Section 5.3.2) is that the MLR model is simple. It does not add variables whose variability is described by previously entered variables or that are not linearly related to the analyte of interest (e.g., have large contributions from interfering species). With the full-spectrum techniques, all sources of variation are implicitly accounted for in the model. This is a more complicated way of dealing with the variation not related to the analyte of interest.

One disadvantage of stepwise MLR over the full-spectrum techniques is that the ability to detect unusual samples is limited because of the elimination of variables. A second disadvantage is that the benefits of multivariate signal aver-aging are largely eliminated (see Section 5.3.2).

5.3.1.1 MLR EXAMPLE 1 For this example, a simple linear two-compo-nent system is considered. The spectra of pure components A and B shown in Figure 5.64 are provided here for learning value and are not required for the analysis. Twenty-five calibration standards are prepared according to a five-level, two-factor full factorial design with the concentrations of A and B ranging from 1 to 5. Twenty-five validation samples are prepared with randomly varying concentrations of A and B within the calibration range. The primary focus of the following discussion is in modeling component A, with limited discussion about the component B model. The discussion follows the "Six Habits of an Effective Chemometrican" which are detailed in Chapter 1.

Habit 1. Examine the Data
The first step is to plot the calibration and validation data (see Figures 5.65*a* and *b* respectively). The spectra look reasonable knowing the pure spectra. If the pure spectra are not known, other chemical knowledge is used to assess the quality of the measurements.

Habit 2. Preprocess as Needed
No preprocessing appears to be needed given the initial observations of the data.

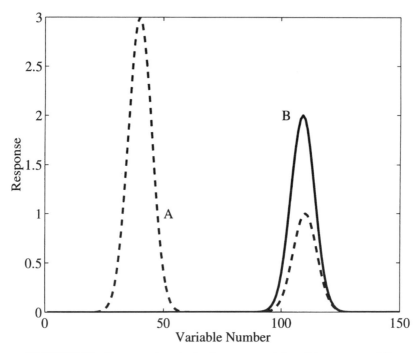

FIGURE 5.64. Pure spectra: dashed, component A; solid, component B.

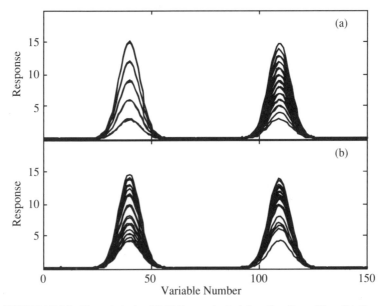

FIGURE 5.65. Raw data for MLR Example 1: (*a*) calibration; (*b*) validation.

Habit 3. Estimate the Model

A mixed variable-selection method is used to select variables with the "Probability to Enter" and "Probability to Leave" values set to 0.05. Separate models are constructed to predict components A and B. The calibration data are used to select the variables to include in the model, and the validation data are used to further refine the model to optimize the predictive ability.

Habit 4. Examine the Results/Validate the Model

Several diagnostic tools are discussed below and a summary is found at the end of the section in Table 5.11. These tools are used to investigate three aspects of the data set: the model, the samples, and the variables. The headings for each tool indicate the aspects that are studied with that tool. The primary use of the model diagnostic tools is to investigate the reasonableness and performance of the model. The sample and variable diagnostic tools are used to identify unusual samples and variables. Other diagnostic tools may be available, depending on the software package employed.

Model and Parameter Statistics (Model Diagnostic): Table 5.9 displays the variables and corresponding statistics for a model constructed to predict component A. The table lists summary statistics for the final regression model and information about the regression coefficients. The Rsquare term (R^2) quantifies the fraction of the variation that is described by the model; Rsquare Adj is the R^2 after it has been adjusted to take into account the number of parameters used to build the model; Root Mean Square Error is an estimate of how well the model fits the concentrations and represents one standard deviation. The rest of Table 5.9 displays the individual variable information for the preliminary subset of variables. Column one lists the order in which the variables are included (the variable deletion steps are not shown). Column two identifies the variable that was added, columns three and four list the corresponding estimated regression coefficient and uncertainties, column five is the t ratio, and column six is the probability of obtaining a t ratio by chance. The t ratio is a measure of how many standard deviations the estimated coefficient is from zero. A large value for the t ratio indicates a significant coefficient and this corresponds to a small Prob $> |t|$ value (Draper and Smith, 1981). A general rule of thumb is to deem a coefficient significant when the Prob $> |t|$ is less than or equal to 0.05.

According to the last column in Table 5.9, the stepwise approach has resulted in a model with 22 variables plus an intercept that are statistically significant at greater than the 95% confidence level. Considering that there are a small number of sources of variation in the mixtures (e.g., only two components) and that the data are known to behave linearly, a 22-variable model is probably excessive.

One reason for the inclusion of a large number of variables is that typical analytical chemistry data do not obey some of the assumptions of the statistical methods used in variable selection (Martens and Næs, 1989). The results from these algorithms can therefore only be used as a guide and should not be

TABLE 5.9. Variable Selection Results for Component A Using the Mixed-Variable Selection Approach

	RSquare	1.00
	RSquare Adj	1.00
	Root Mean Square Error	1.0×10^{-6}

Order of Inclusion	Variable Added	Coefficient Estimate	Standard Error of Coefficient	t Ratio	Prob $> |t|$
1	Intercept	1.82E − 02	8.00E − 06	2.40E + 03	0.0000
2	39	1.41E − 01	1.20E − 05	1.15E + 04	0.0000
3	45	1.41E − 01	2.60E − 05	5.49E + 03	0.0000
4	131	1.50E − 01	1.70E − 05	8.64E + 03	0.0000
5	60	−5.03E − 02	1.30E − 05	−3.90E + 03	0.0000
6	36	1.55E − 01	9.00E − 06	1.81E + 04	0.0066
7	3	1.34E − 01	1.70E − 05	7.99E + 03	0.0009
8	139	1.23E − 01	1.90E − 05	6.64E + 03	0.0000
9	72	−1.06E − 01	2.00E − 05	−5.40E + 03	0.0000
10	149	1.81E − 02	2.50E − 05	7.21E + 02	0.0000
11	2	−1.02E − 01	2.60E − 05	−3.90E + 03	0.0000
12	87	−4.74E − 02	1.40E − 05	−3.40E + 03	0.0000
13	11	−4.60E − 02	2.60E − 05	−1.76E + 03	0.0000
14	62	−2.03E − 02	2.40E − 05	−8.56E + 02	0.0001
15	84	−2.04E − 02	1.60E − 05	−1.30E + 03	0.0000
16	130	−2.59E − 02	3.10E − 05	−8.25E + 02	0.0000
17	6	−2.15E − 02	4.80E − 05	−4.51E + 02	0.0000
18	32	−1.50E − 02	3.40E − 05	−4.41E + 02	0.0002
19	147	−1.28E − 02	3.40E − 05	−3.79E + 02	0.0000
20	69	3.31E − 03	2.90E − 05	1.13E + 02	0.0000
21	124	1.45E − 03	1.90E − 05	7.80E + 01	0.0000
22	28	−6.45E − 04	1.90E − 05	−3.38E + 01	0.0000
23	18	−2.41E − 04	2.00E − 05	−1.22E + 01	0.0000

considered the final judge of which variables are used. Variables should only be included if they are reasonable from a chemical and/or physical perspective (Callis, 1988).

To illustrate the problem with the statistical significance of the variables listed in Table 5.9, consider generating 25 random numbers, one for each sample in the calibration set. For this new variable to appear to be a "good" variable for fitting the concentration of component A, a linear relationship must exist between the 25 random numbers and the concentration of component A in the samples. The likelihood of this is small given one set of random numbers. However, given multiple sets of 25, eventually a good linear relationship will be found. When there are many variables that contain just random noise, this is like generating many sets of random numbers. The result is that variable selection techniques typically select too many variables as being statistically significant when applied to highly correlated data like that encountered in spectroscopic applications. (Tenge, 1989; Freedman, 1983; Rencher and Pun, 1980).

The problem with keeping variables that appear to be significant but are only modeling noise is that this "overfitting" of the data degrades the prediction ability of the model. It is, therefore, important to only add variables that improve prediction of future samples, not just improve the fit. The approach we take in this section is to use the statistical output as the first pass for variable selection. We then further refine the model (which usually means reducing the number of variables) by examining results from a validation set.

Root Mean Square Error of Calibration (RMSEC) Plot (Model Diagnostic): As a summary of how well models with different numbers of variables fit the calibration data, the root mean square error of calibration (RMSEC) is calculated as the number of variables is varied using Equation 5.26:

$$\text{RMSEC} = \sqrt{\frac{\sum_{i=1}^{\text{nsamp}} (c_{i,\text{cal}} - \hat{c}_{i,\text{cal}})^2}{\text{nsamp} - (\text{nsel} + 1)}} \qquad (5.26)$$

where $c_{i,\text{cal}}$ is the concentration of the ith calibration sample, $\hat{c}_{i,\text{cal}}$ is the fitted concentration for the ith sample, using a model with nsel variables, and nsamp is the number of samples in the calibration set. The numerator measures the error in fitting the concentration values and always decreases as variables are added to the model. The denominator is a degrees of freedom term (nsel + 1 is subtracted from nsamp because the data were used to estimate this number of coefficients). Because the denominator may get smaller faster than the numerator, it is possible for the RMSEC to go through a minimum as variables are added. Typically, however, the curve continually decreases as statistically significant variables are added. Note also that for the same number of variables in the model, the root mean square error output from the statistical packages (e.g., Table 5.9) is equal to RMSEC.

For the example data, the RMSEC is calculated for models containing 1–22 variables (adding the variables in the order listed in Table 5.9). The RMSEC versus the number of variables included in the model is plotted in Figure 5.66 for component A. The fit improves as variables are added to the model (RMSEC decreases). However, knowing these results reflect model fit, there is a concern about overfitting (i.e., fitting noise from the calibration data). It is known that the error in the concentration values is 0.010 (1σ). The RMSEC drops below this level after the fifth variable is included, and therefore the tentative conclusion is that a four-variable model is appropriate.

Root Mean Square Error of Prediction (RMSEP) Plot (Model Diagnostic): The number of variables to include is finalized using a validation procedure that accounts for predictive ability. There are two approaches for calculating the prediction error: internal cross-validation (e.g., leave-one-out cross-validation with the calibration data) or external validation (i.e., perform prediction

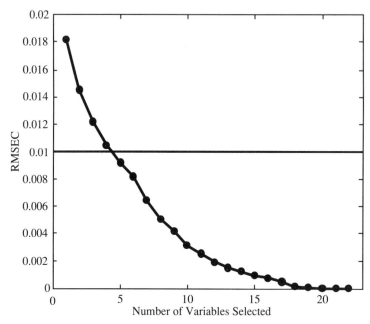

FIGURE 5.66. RMSEC versus number of variables in the model for component A. The horizontal line represents the error in the known concentrations.

on a separate validation set). The latter approach is used for this example and all of the diagnostics that follow are applied to the validation data (except Cook's distance). The figure of merit that is evaluated is the root mean square error of prediction (RMSEP), which is calculated as follows:

$$\text{RMSEP} = \sqrt{\frac{\sum_{i=1}^{\text{nsamp}} (c_{i,\text{val}} - \hat{c}_{i,val})^2}{\text{nsamp}}} \tag{5.27}$$

where $c_{i,\text{val}}$ is the concentration of the ith validation sample, $\hat{c}_{i,\text{val}}$ is the predicted concentration for the ith sample using a model with nsel variables, and nsamp is the number of samples in the validation set. With RMSEP the number of degrees of freedom is nsamp because the validation data are not used to estimate the regression coefficients. As with RMSEC, a small value for RMSEP is desirable, but now the statistic is measuring predictive ability versus fit.

The RMSEP plotted versus the number of variables included in the model for component A indicates that the optimum model contains one variable (see Figure 5.67). The prediction error with one variable (0.016) is greater than the known concentration error of 0.010, which is an indication that the model is not overfitting. Assuming the model is correct, the fact that the RMSEP is greater than the errors in the known concentrations is due to errors in **R**.

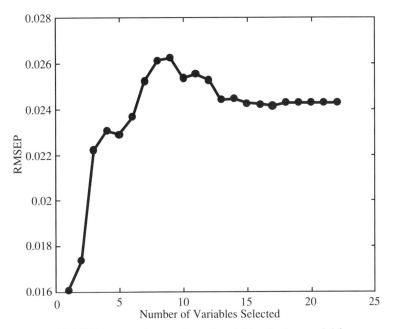

FIGURE 5.67. RMSEP versus the number of variables in the model for component A using a separate validation set.

The RMSEC in Table 5.10 (0.018) is larger than the RMSEP (0.016). This seems incorrect, because prediction errors are typically larger than fit errors, but these numbers are not statistically different.

The next step is to construct the regression model from the calibration data using one variable plus an intercept. The results of this (now univariate) model (see Table 5.10) indicate that the intercept is not a significant term in the regression. This is expected because the data are known to obey Beer's Law. Traditionally, in statistics the intercept term is left in the model even when found to be nonsignificant.

TABLE 5.10. MLR Model for Component A Using One Variable Plus an Intercept

		RSquare	0.9998		
		RSquare Adj	0.9998		
		Root Mean Square Error	0.018		

Variable Number	Variable Added	Coefficient Estimate	Standard Error of Coefficient	t Ratio	Prob $> |t|$
1	Intercept	0.013	0.0085	1.51	0.1451
2	39	0.339	0.0009	389.55	0.0000

The RMSEP for component B levels off after two variables at 0.041 (not shown), and has a corresponding RMSEC = 0.032 and RSquare adjusted = 0.999. The errors are larger than the known concentration error of 0.010, indicating that the model is not overfitting. Assuming the model is correct, the fact that the RMSEP is greater than the errors in the known concentrations is due to errors in **R**. Again, the intercept is not significant, but is kept in the model.

Selected Variable Plot (Model and Variable Diagnostic): The variables that have been included in the model should be examined to see if they are reasonable given knowledge about the chemistry of the samples and measurements. Figure 5.68 displays the calibration spectra with vertical lines indicating the variables selected for the models for components A and B.

The variable chosen for component A is near the peak at variable 40. No variables were selected from the peak near variable 110. Figure 5.64 shows that the pure spectrum for component B overlaps with the component A peak in the 110 region. It is, therefore, reasonable that this region is excluded from the model for predicting component A. This is another example of how the inverse model can ignore regions with interfering signals.

The two-variable model for component B includes a variable from both of the prominent features in the spectra. Variable 112 is included to model component B and variable 33 is included to compensate for the interference of A.

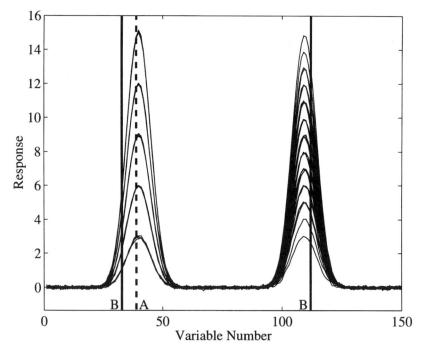

FIGURE 5.68. Calibration spectra with the one wavelength selected for modeling component A (dashed vertical line) and the two wavelengths selected for modeling component B (solid vertical lines).

Cook's Distance Plot (Model Diagnostic): A statistic known as Cook's distance can be used to detect calibration data outliers by identifying which samples are most influential on the model. Now that the selected variables have been finalized, it is good practice to examine the calibration data for influential samples. These samples should be investigated and removed if it is determined that they have an unusual effect on the model.

To determine the influence, the regression parameters are estimated with and without a sample in the model. If the parameters change significantly when the sample is removed, it will have a large Cook's distance value. A Cook's distance greater than 1 suggests a potentially important change in the model parameters when the sample is removed (Weisberg, 1985). In Figure 5.69 Cook's distance is plotted versus sample number for the component A calibration data using a one-variable model. All of the distances are significantly less than 1 and, therefore, none of the calibration samples are suspect outliers.

Predicted vs. Known Concentration Plot (Model and Sample Diagnostic): Returning to the validation data, Figure 5.70 shows the predicted versus known concentrations for component A using the one-variable model. For model diagnostics, this plot is examined for bias or lack of fit by a large number of samples. This would indicate an inability of the model to predict component A. See also Section 5.2.1.1 for more discussion of this diagnostic. All of the samples in Figure 5.70 appear to be clustered about the ideal line, which indicates a good model with no outliers.

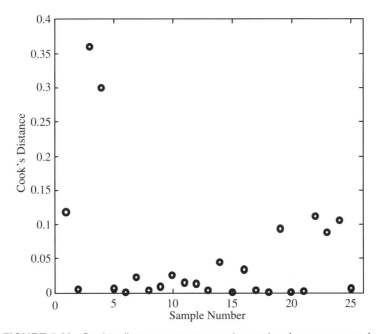

FIGURE 5.69. Cook's distance versus sample number for component A.

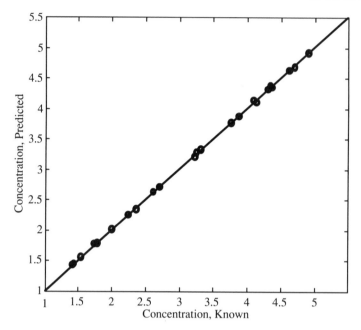

FIGURE 5.70. Predicted concentration versus known concentration of component A for the validation data using a one-variable model.

Concentration Residuals versus Predicted Concentration Plot (Model and Sample Diagnostic): Figure 5.71 is a plot of the concentration residuals versus the predicted values for component A. (See Section 5.2.1.1 for more discussion on how to interpret this plot.) The residuals vary from -0.035 to $+0.020$ and no unusual features or patterns are observed.

Concentration Residuals vs. Other Postulated Variable Plot (Model and Sample Diagnostic): The concentration residuals in this example are acceptable, and so it is assumed that all relevant sources of variation are being modeled. No "other postulated parameters" are hypothesized.

Statistical Prediction Error vs. Sample Number Plot (Sample Diagnostic): A statistic is available for the predicted values using Equation 5.28. We will refer to this as the statistical prediction error to distinguish it from the observed concentration residuals.

$$\text{statistical prediction error} = s\sqrt{\mathbf{r}_0^\mathrm{T}(\mathbf{R}^\mathrm{T}\mathbf{R})^{-1}\mathbf{r}_0} \qquad (5.28)$$

In Equation 5.28, s is a function of the concentration residuals observed during calibration, $\mathbf{r}_0$ is the measurement vector for the prediction sample, and $\mathbf{R}$ contains the calibration measurements for the variables used in the model. Because the assumptions of linear regression are often not rigorously obeyed, the statistical prediction error should be used empirically rather than absolutely. It is useful for validating the prediction samples by comparing the values for

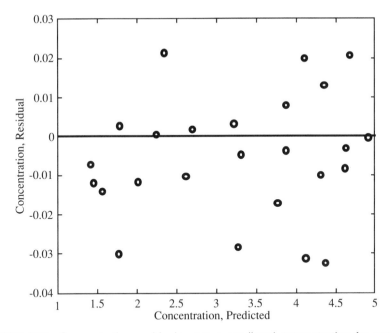

FIGURE 5.71. Concentration residuals versus predicted concentration for component A for the validation data using a one-variable model.

the unknown samples to those found from the validation phase (see Habit 6). The statistical prediction errors for the validation samples are plotted in Figure 5.72 for component A. The maximum value is 6.1×10^{-3} for component A and 1.2×10^{-2} for component B (not shown). No validation samples appear to have unusual values for either the component A or component B models.

Summary of Validation Diagnostic Tools for MLR, Example 1: The MLR method with variable selection was used to construct a model to predict the concentration of components A and B in mixtures. The model for component A is discussed in depth. The details of the data set used to construct the calibration model are as follows:

Calibration design: 25 samples, 5-level full factorial design

Validation design: 25 samples with randomly varying concentrations

Preprocessing: none

Variable range: 150 original measurement variables

The operating ranges of the prediction model based on the calibration data follow. Predicting future samples from outside this operating range is extrapolating and may produce unreliable results.

Components A and B: 1–5 units

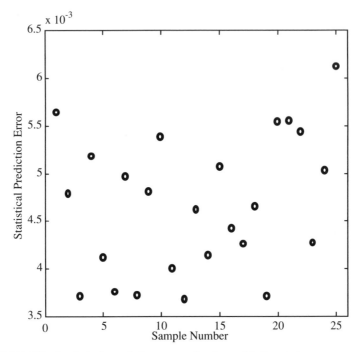

FIGURE 5.72. Statistical prediction errors for the validation data, component A.

Candidate variables were chosen using a mixed-variable selection method and validated based on prediction ability. Separate models (with different measurement variables) were estimated for each of the components. The final models and measures of performance are as follows (see Table 5.11 for a description of these figures of merit):

Probability to enter and leave = 0.05
Component A model
    Variable(s): No. 39, plus intercept
    Adjusted R^2 = 0.999
    RMSEC = 0.018
    RMSEP = 0.016
    Maximum statistical prediction error = 0.006
    Expected error in prediction = ±0.03
    (Based on errors in predicted A concentration. See Figure 5.71.)
Component B model:
    Variable(s): Nos. 33 and 112, plus intercept
    Adjusted R^2 = 0.999
    RMSEC = 0.032
    RMSEP = 0.041
    Maximum statistical prediction error = 0.012

TABLE 5.11. Summary of Validation Diagnostics for MLR

Diagnostics	Description and Use
Model	
Model and parameter statistics	Model statistics include R^2, adjusted R^2 and root mean squared error. Parameter statistics are the estimated regression coefficients and associated statistics.
	The statistics are used to evaluate the quality of the model.
	Acceptable models have R^2 and adjusted R^2 values near 1 and a root mean squared error that is comparable to the known errors. Only statistically significant regression coefficients are included in the model.
	Many of the regression assumptions are not obeyed when analyzing chemical data (e.g., spectroscopic data). In these cases the statistical output can be misleading and should only be used for qualitative assessment.
Root mean square error of calibration (RMSEC) plot (RMSEC vs. number of variables)	A single number that quantifies the fit of the model to the data for a given number of variables included in the model.
	The ideal RMSEC plot decreases as variables are added into the model and levels off or increases when it reaches a value that is comparable to the known concentration errors.
	The optimum number of variables to include is the number of variables at which the RMSEC levels off.
Root mean square error of prediction (RMSEP) plot (RMSEP vs. number of variables)	A single value that quantifies the magnitude of the concentration residuals for the validation samples for a given number of variables included in the model.
	Is used to help determine the optimal number of variables to include into the model.
	The ideal RMSEP plot decreases as variables are added into the model and increases when it reaches a value that is comparable to the known concentration errors. The optimum number of variables to include is the number of variables after which the RMSEP increases.
	The RMSEP is a more realistic estimate of predictive ability than RMSEC.
Selected variable plot	Plot of the measurement vectors with vertical lines indicating the variables that have been included in the model.
	Used to verify that the selected variables are reasonable based on knowledge of the chemistry and measurement system.
Cook's distance plot (Cook's distance vs. sample number)	A single number that quantifies the influence of each calibration sample on the model.
	Used to identify unusual samples in the calibration set. Investigate samples with Cook's distances greater than 1.

continued

TABLE 5.11. *(Continued)*

Diagnostics	Description and Use
Predicted vs. known concentration plot ($\hat{c}$ vs. c)	Predicted concentrations of validation samples plotted versus the known values.
	Used to determine whether the model accounts for the concentration variation in the validation set.
	Points are expected to fall on a straight line with a slope of one and a zero intercept.
	A modeling problem is indicated when a systematic pattern away from the ideal line is observed.
Concentration residual vs. predicted concentration plot [$(c - \hat{c})$ vs. $\hat{c}$]	The differences between the known and predicted concentrations (residuals) are plotted versus the predicted concentrations for validation samples.
	Used to determine whether the model accounts for the concentration variation in the validation set.
	Points are expected to vary randomly about a line of zero slope and zero intercept.
	A modeling problem is indicated when a systematic pattern away from the ideal line is observed.
	Patterns are often more discernible in this plot compared to the predicted versus known plot.
Concentration residuals vs. other postulated variable plot [$(c - \hat{c})$ vs. x]	The concentration residuals can be plotted versus a postulated variable such as run order.
	Used to determine if the model is accounting for the variability correlated with the postulated variable.
	Points are expected to vary randomly about a line of zero slope and zero intercept.
	A modeling problem is indicated when a systematic pattern away from the ideal line is observed.
	Identified effects should be controlled or added to the model if possible.
Sample	
Predicted vs. known concentration plot ($\hat{c}$ vs. c)	Predicted concentrations of validation samples plotted versus the known values.
	This plot can be used to identify outlier samples.
	Points are expected to fall on a straight line with a slope of one and a zero intercept.
	An outlier is indicated when a point is far from the ideal line relative to the other points.
Concentration residual vs. predicted concentration plot [$(c - \hat{c})$ vs. $\hat{c}$]	The differences between the known and predicted concentrations (residuals) are plotted versus the predicted concentrations for the validation samples.
	This plot can be used to identify outlier samples.
	Points are expected to vary randomly about a line of zero slope and zero intercept.
	A small number of samples with unusual concentration residuals is an indication of sample rather than model problems.

TABLE 5.11. *(Continued)*

Diagnostics	Description and Use
	Outliers are often more discernible in this plot compared to the predicted versus known plot.
Statistical prediction error vs. sample number plot (statistical prediction error vs. sample number)	The statistical prediction errors are the uncertainty in the predicted concentrations for each validation sample.
	It is an optimistic estimate of the concentration error for each sample because it only includes precision.
	Statistical prediction errors for a sample that are quite different than the other validation samples indicate a problem with that sample.
Variable	
Selected variable plot	Plot of the measurement vectors with vertical lines indicating the variables that have been included in the model.
	Used to verify that the selected variables are reasonable based on knowledge of the chemistry and measurement system.

Table 5.11 summarizes the validation diagnostic tools discussed in this section. The first column in the table lists the name of each tool and the second column describes results from both well-behaved and problematic data.

Habit 5. Use the Model for Prediction

The models for predicting components A and B can now be used to predict unknown samples. As shown in Equation 5.25, prediction is the multiplication of the regression vector (the model) by the selected measurement variables of the unknown sample. The predicted values for 25 unknown samples are plotted in Figure 5.73 for components A and B.

Habit 6. Validate the Prediction

Three prediction diagnostic tools are discussed below and a summary is found at the end of the section in Table 5.12. As discussed in the introduction to this section, one disadvantage of models with few variables is that the prediction validation diagnostics are limited.

Predicted Concentration Plot: Figure 5.73 displays the predicted concentrations for components A and B with the horizontal lines representing the range of the concentrations in the calibration data (the range is the same for both components). Sample 2 has a predicted value outside the calibration range for component A. This prediction may not be accurate because the model is being used for extrapolation, which in general is not advisable.

Statistical Prediction Errors: The statistical prediction errors for component A shown in Figure 5.74 reveal that sample 2 has a larger value than the maximum value observed in the validation phase (denoted by the horizontal line). This same sample also has a large statistical prediction error for component B (not shown). These large statistical prediction errors are not surprising, given

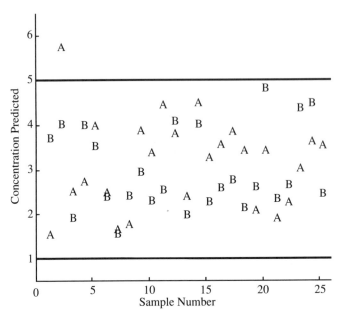

FIGURE 5.73. Predicted concentrations of components A and B for the unknowns with the horizontal line indicating the calibration ranges for both A and B.

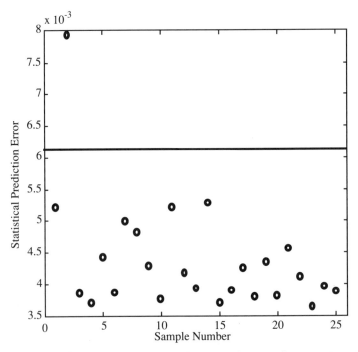

FIGURE 5.74. Statistical prediction errors for the unknowns for component A with the horizontal line indicating the maximum from model validation.

that the predicted value for this sample is outside the range of the calibration data and the statistic is proportional to the magnitude of the measurement vector (see Equation 5.28). Although its concentration for component B is within the calibration range, the statistical prediction error for B is also inflated because the model for component B includes a variable where A responds.

Raw Measurement Plot: In multivariate calibration, it is normally not necessary to plot the prediction data if the outlier detection technique has not flagged the sample as an outlier. However, with MLR, the outlier detection methods are not as robust as with the full-spectrum techniques (e.g., CLS, PLS, PCR) because few variables are considered. Figure 5.75 shows all of the prediction data with the variables used in the modeling noted by vertical lines. One sample appears to be unusual, with an extra peak centered at variable 140. The prediction of this sample might be acceptable because the peak is not located on the variables used for the models. However, it is still suspect because the new peak is not expected and can be an indication of other problems.

Because the data are synthetic, it is known that four additional prediction samples are problematic. Samples 3 and 10 have twice and five times the noise level of the other samples, respectively. Additionally, samples 15 and 20 are shifted by 0.4 and 0.7 units relative to the correct position. To show the

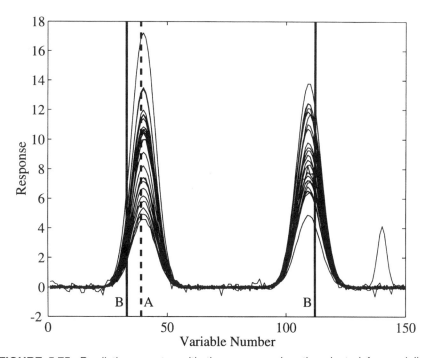

FIGURE 5.75. Prediction spectra with the one wavelength selected for modeling component A (dashed vertical line) and the two wavelength selected for modeling component B (solid vertical lines).

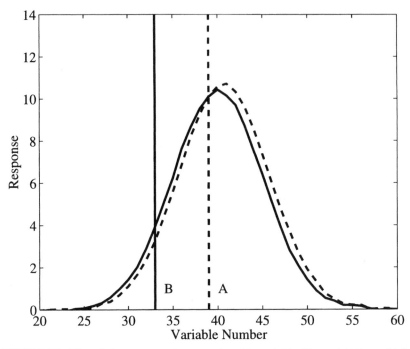

FIGURE 5.76. Plot of the spectra for samples 18 and 20. Normal (18), solid line; shifted (20), dashed line. The dashed vertical line indicates the variable used in the modeling of component A, and the solid vertical line indicates the variable used in the modeling of component B.

magnitude of the shift, a region of the sample 20 spectrum is plotted with a representative normal spectrum in Figure 5.76. The statistical prediction errors did not flag any of the four known problematic samples, which illustrates the limitation in prediction diagnostics when using models with small numbers of variables. In contrast, these samples are detected when using a full-spectrum method.

Summary of Prediction Diagnostic Tools for MLR, Example 1: All but one of the 25 prediction samples are predicted to be within the calibration range, and the outlier statistics indicate that reasonable confidence can be placed in all samples predicted within the calibration range. Based on the range of validation concentration residuals, the errors in the predicted component A concentrations of the unknowns are expected to be within ±0.03 concentration units. However, four additional unusual samples were not detected, which demonstrates the limitations of the outlier diagnostics.

Table 5.12 summarizes the prediction diagnostic tools discussed in this section. The first column in the table lists the name of each tool and the second column discusses what is expected with well-behaved and problematic data.

TABLE 5.12. Summary of Prediction Diagnostics for MLR

Prediction Diagnostics	Description and Use
Predicted concentration plot ($\hat{c}$ vs. sample number)	A graphical means to display predicted concentrations along with the calibration range.
	Used to help assess the reliability of the predicted concentrations of unknown samples.
	The predicted concentrations should fall within the calibration concentration range.
Statistical prediction errors (statistical prediction error vs. sample number)	The statistical prediction errors are the uncertainty in the predicted concentrations for each unknown sample.
	It is an optimistic estimate of the concentration error for each unknown sample because it only includes precision.
	Used to help assess the reliability of the predicted concentrations of unknown samples.
	Statistical prediction errors for a sample that are quite different than the validation samples indicate a problem with that unknown sample.
Raw measurement plot ($\mathbf{r}$ vs. variable number)	A plot of the measurement vectors for the unknown samples with vertical lines indicating the variables that have been included in the model.
	Used to investigate unknown samples that have been identified as being unusual.
	Used in conjunction with knowledge of the chemistry and the measurement system to assign cause when suspect predictions are identified.

5.3.1.2 MLR EXAMPLE 2 In this second example, the determination of caustic concentration in an aqueous sample containing NaOH (caustic) and salt using NIR spectroscopy is examined. This example is discussed in Section 5.2.2.2 as well as in Section 5.3.2.2 where PLS is applied.

Habit 1. Examine the Data
The examination of the raw data is discussed in Section 5.2.2.2.

Habit 2. Preprocess as Needed
The preprocessing is also discussed in Section 5.2.2.2. Only variables between 9087 and 8315 cm^{-1} are considered in this analysis. A first derivative is taken before variable selection.

Habit 3. Estimate the Model
The mixed-variable selection method is used with the "probability to enter" and "probability to leave" values set to 0.05. Only a model for the prediction of caustic concentration is developed, which means that the values for the salt, water, and temperatures are not used in the calculations (even though in this case they are known). The calibration data ($\times$ in Figure 5.42) consisting

of 7 design points (samples) and 62 spectra are used to estimate the variables to include in the model, and the validation data ($\bigcirc$ in Figure 5.42) consisting of 5 design points and 33 spectra are used to further refine the model to optimize the predictive ability. Multiple spectra collected at varying temperatures were obtained for each design point.

Habit 4. Examine the Results/Validate the Model

The discussion below follows the validation diagnostics found in Table 5.11, Section 5.3.1.1.

Model and Parameter Statistics (Model Diagnostic): Table 5.13 displays the variables selected for a model constructed to predict caustic. The table lists summary statistics for the regression model as well as information about the estimated regression coefficients. Six variables in addition to an intercept are found to be significant at the 95% confidence level.

Root Mean Square Error of Calibration (RMSEC) Plot (Model Diagnostic): The RMSEC as a function of the number of variables included in the model is shown in Figure 5.77. It decreases as variables are added to the model and the largest decrease is observed between a one- and two-variable model. The reported error in the reference caustic concentration is approximately 0.033 wt.% (1σ). The tentative conclusion is that four variables are appropriate because the RMSEC is less than the reference concentration error after five variables are included in the model.

Root Mean Square Error of Prediction (RMSEP) Plot (Model Diagnostic): The validation set is employed to determine the optimum number of variables to use in the model based on prediction (RMSEP) rather than fit (RMSEC). RMSEP as a function of the number of variables is plotted in Figure 5.78 for the prediction of the caustic concentration in the validation set. The curve levels off after three variables and the RMSEP for this model is 0.053 wt.%. This value is within the requirements of the application ($1\sigma = 0.1$) and is not less than the error in the reported concentrations.

TABLE 5.13. Variable Selection Results for Caustic Using a Mixed-Variable Selection Approach

		RSquare	0.9997		
		RSquare Adj	0.9997		
		Root Mean Square Error	0.0233		

Order of Inclusion	Wave-number	Coefficient Estimate	Standard Error of Coefficient	t Ratio	Prob $> |t|$
1	Intercept	42.1	1.5	27.4	0.0000
2	8439	2399.6	424.0	5.7	0.0000
3	8709	2101.7	279.1	7.5	0.0000
4	8609	−2684.5	178.2	−15.1	0.0000
5	8917	7312.2	556.7	13.1	0.0000
6	8747	−5098.5	288.9	−17.6	0.0000
7	8354	−2090.9	619.2	−3.4	0.0014

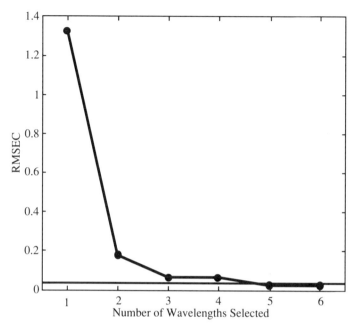

FIGURE 5.77. RMSEC as a function of the number of variables for MLR Example 2. The horizontal line represents the error in the known caustic concentrations.

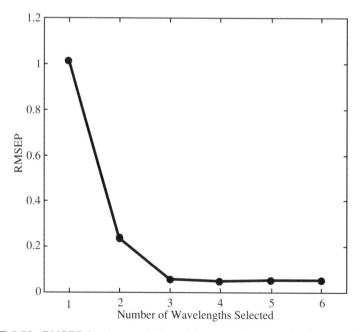

FIGURE 5.78. RMSEP for the prediction of the caustic concentration as a function of the number of variables using a separate validation set.

TABLE 5.14. MLR Model for Caustic Using Three Variables Plus an Intercept

		RSquare	0.9984		
		RSquare Adj	0.9983		
		Root Mean Square Error	0.0608		

Variable Number	Wave-number (cm^{-1})	Coefficient Estimate	Standard Error of Coefficient	t Ratio	Prob > |t|
1	Intercept	50.0	0.2	204.8	0.0000
2	8439	−2561.2	61.2	−41.9	0.0000
3	8709	−2660.1	16.4	−162.2	0.0000
4	8609	−608.9	28.9	−21.0	0.0000

The next step is to construct a model with the calibration data using three variables plus an intercept. The results of this model (Table 5.14) are that Rsquare Adj is greater than 0.998 and all of the coefficients in the model are significant at greater than a 95% confidence level.

Selected Variable Plot (Model and Variable Diagnostic): Figure 5.79 shows the three selected variables for the first derivative preprocessed data.

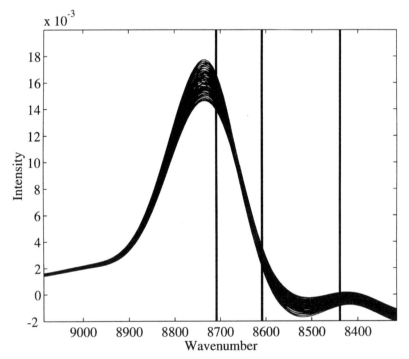

FIGURE 5.79. Preprocessed calibration spectra with the three wavelengths selected for predicting the caustic concentration.

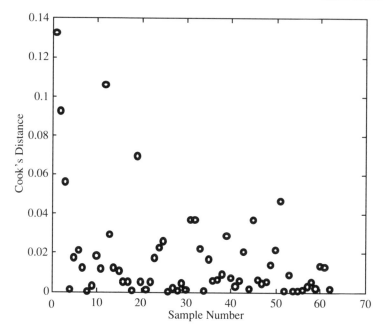

FIGURE 5.80. Cook's distance versus sample number for the three-component model.

Extensive interpretation is not possible because the effects of the salt, caustic, and temperature are manifested as complex perturbations on the water band.

Cook's Distance Plot (Model Diagnostic): In Figure 5.80, the Cook's distance is plotted versus sample number for the three-variable caustic model. All values are much less than 1 (the maximum is 0.14), which is an indication that no samples are potentially overly influential on the model parameters.

Predicted vs. Known Concentration Plot (Model and Sample Diagnostic): The predicted versus known concentrations for the validation samples using the three-variable model are shown in Figure 5.81. All the points are clustered close to the ideal line, indicating that the model is predicting these samples well. No samples need to be investigated further because none are unusually far from the line.

Concentration Residual vs. Predicted Concentration Plot (Model and Sample Diagnostic): The concentration residuals plotted as a function of the predicted concentrations are shown in Figure 5.82. There is some indication of curvature, but this is difficult to assess with only four concentration levels. The predictions are all within ±0.1 wt.%, which is well within the specification of the application. There do not appear to be any problems with individual samples.

Concentration Residuals vs. Other Postulated Variable Plot (Model and Sample Diagnostic): The concentration residuals are plotted as a function of the sample temperature in Figure 5.83. This is examined to verify that the model adequately accounts for the temperature effect on the spectra. The conclusion is that the model has adequately accounted for this influence on the

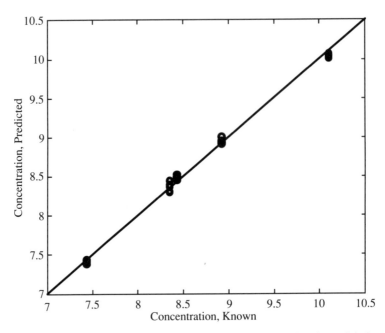

FIGURE 5.81. Predicted versus known caustic concentration for the validation samples, three-variable MLR model.

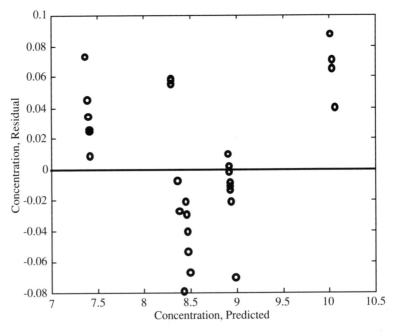

FIGURE 5.82. Concentration residuals versus predicted caustic concentration for the validation samples, three-variable MLR model.

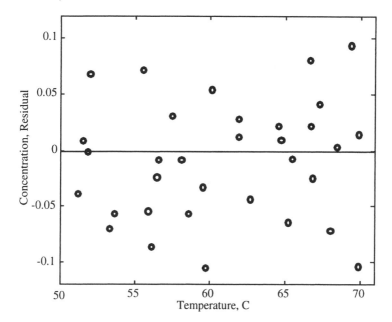

FIGURE 5.83. Concentration residuals as a function of temperature for the validation samples.

spectra because there is no structure in the concentration residuals as a function of temperature. This has been accomplished even though the temperature measurements have not been included in the calculations (as compared to ICLS, Section 5.2.2.2).

Statistical Prediction Error vs. Sample Number Plot (Sample Diagnostic): The statistical prediction errors for the validation data are shown in Figure 5.84. There are no samples which have an error that is unusual relative to the rest of the validation data. This further confirms the earlier conclusion that there are no outlier samples. The maximum of 0.029 will be used for assessing the reliability of prediction in Habit 6.

Summary of Validation Diagnostic Tools for MLR, Example 2: The determination of caustic concentration in an aqueous sample containing NaOH (caustic) and salt using NIR spectroscopy is examined in this example. The goal was to create a model to predict caustic concentration in aqueous samples with varying salt and temperature levels. The details of the data set used to construct the calibration model are as follows:

Calibration Design: 62 spectra, 7 design points
(collected with varying temperature)

Validation Design: 33 spectra, 5 design points
(collected with varying temperature)

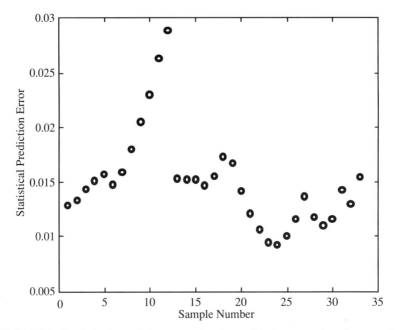

FIGURE 5.84. Statistical prediction error for the validation samples, three-variable MLR model.

Preprocessing:	15-point first derivative
Variable Range:	101 original measurement variables, from 9087–8315 cm^{-1}

The operating ranges of the prediction model based on the calibration data set follow. Predicting future samples from outside this operating range is extrapolating and may produce unreliable results.

Caustic: 6.4–10.4 wt.%
Salt: 13.0–17.0 wt.%
Temperature: 50.2–69.8°C

Candidate variables were chosen using a mixed-variable selection method and validated based on prediction ability. The final model details for caustic and measures of performance are as follows (see Table 5.11 for a description of these figures of merit):

Probability to enter and leave = 0.05
Variable(s): 8439, 8709, 8609 cm^{-1}, plus intercept
Adjusted R^2 = 0.998
RMSEC = 0.0608 wt.%

RMSEP = 0.053 wt.%

Maximum statistical prediction error = 0.029 wt.%

Expected error in prediction = ±0.1 wt.%

(Based on errors in predicted caustic concentration. See Figure 5.82.)

This is an acceptable model performance based on the requirements of the application. The inverse model accounts for the effects of salt and temperature on the water peak without explicitly using the salt or temperature information in the calculations. This is in contrast to the ICLS analysis of these same data where the concentrations and temperatures of the calibration data are required to obtain a satisfactory model (Section 5.2.2.2).

Habit 5. Use the Model for Prediction
The predicted caustic concentrations for 99 prediction samples are plotted in Figure 5.85. The first derivative over the entire wavelength region is calculated before performing a prediction using the three-variable model.

Habit 6. Validate the Prediction
This discussion follows the prediction diagnostics found above in Table 5.12, Section 5.3.1.1.

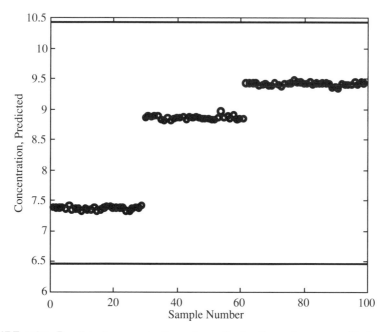

FIGURE 5.85. Predicted concentration of caustic for the unknowns, with the horizontal lines indicating the range of calibration samples.

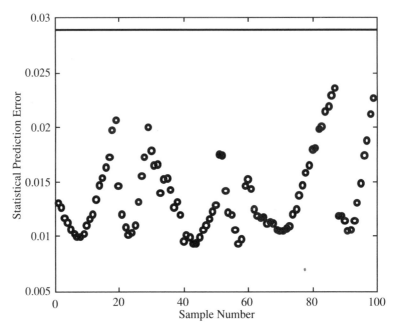

FIGURE 5.86. Statistical prediction errors for the caustic predictions, with the horizontal line indicating the maximum from calibration.

Predicted Concentration Plot: All predictions are within the calibration range, indicating that the model is not extrapolating (see Figure 5.85).

Statistical Prediction Errors: The statistical prediction errors are plotted in Figure 5.86, with the maximum from the model validation denoted by the horizontal line. All prediction samples fall below this line, indicating that there are no unusual samples.

Raw Measurement Plot: The raw data for the prediction samples is not usually plotted given that the diagnostics did not flag any samples as unusual. They are displayed here because of the limited error detection of the other diagnostic tools. The preprocessed data in Figure 5.87 do not reveal any unusual samples.

Summary of Prediction Diagnostic Tools for MLR, Example 2: The conclusion is that the predicted concentrations for all 99 unknown samples are reliable. Based on the range of validation concentration residuals (see Figure 5.82), the errors in the predicted caustic concentrations of the unknowns are expected to be within ± 0.10 wt.%.

5.3.2 Partial Least Squares/Principal Component Regression

Partial least squares (PLS) and principal component regression (PCR) are the most widely used multivariate calibration methods in chemometrics. Both of these methods make use of the inverse calibration approach, where it is

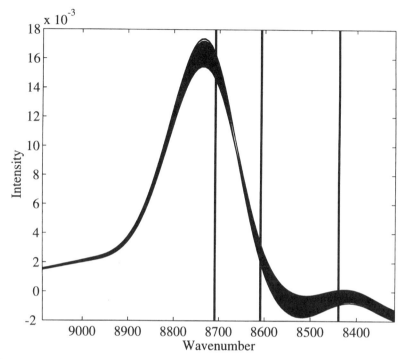

FIGURE 5.87. Preprocessed prediction samples, with the vertical lines indicating the variables used in the MLR model.

possible to calibrate for the desired component(s) while implicitly modeling (accounting for) the other sources of variation. Estimation of the inverse calibration model involves the inversion of a typically unstable matrix. Both PLS and PCR solve this inversion problem by replacing the original variables with linear combinations of the variables (factors). The difference between PLS and PCR is in how the factors are calculated.

Given the two approaches for solving the inverse calibration problem, MLR with variable selection or these factor-based methods, what are the advantages of the "full-spectrum" methods (PCR, PLS) over MLR? Using stepwise MLR, only a few variables are selected to construct the model. Although this leads to a simple model, there are two advantages to using more variables. First, the models become more sensitive to the detection of samples that are unusual compared to the calibration set (outliers). This difference in error detection is demonstrated in Figure 5.88, where 100 measurements are made on two unknown samples. Assume that the dashed spectrum is similar to the calibration data (not shown here) and the solid spectrum is unusual. Further assume that the vertical lines on the graph indicate two variables that were selected for a stepwise MLR model. In this example, the MLR prediction diagnostics would not flag the unusual sample. In fact, because the two samples have the same intensities at the chosen variables, MLR would predict that these two samples have the same concentration. In contrast, the full-spectrum techniques would

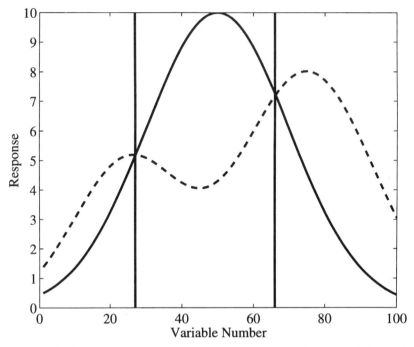

FIGURE 5.88. Demonstration of the error detection capabilities of the full-spectrum methods. The dashed line represents a spectrum that has a similar shape to the calibration data. The solid-line spectrum has a quite different shape than the calibration spectra. The vertical lines are two variables that have been selected for MLR.

easily detect that the one unknown sample has a different shape than the calibration samples, indicating an unreliable prediction result.

A second advantage of the full-spectrum techniques over the stepwise MLR approach is the "multivariate advantage." This is related to signal averaging, where the standard deviation of a measurement is reduced by a factor of $\sqrt{n}$ when the average of n determinations is used in place of a single measurement. With multivariate data, signal averaging takes place when many nearly redundant measurements are used to construct a model. An analogy familiar to many chemists is encountered in the field of chromatography. It is common practice to relate peak area to concentration, as opposed to simply measuring the peak height at a given retention time. This is because the precision of the area is usually much better than the precision of the intensity at a given point.

Mathematical Description of PLS and PCR: In this section, the PCR algorithm is described as a two-step procedure of PCA followed by MLR. Although in practice the steps are combined, we feel this is the most intuitive approach to understanding the algorithm. This description of PCR is followed by a brief discussion of the differences between PLS and PCR.

Recall the form of the inverse model is $\mathbf{c} = \mathbf{Rb}$ (see Equation 5.19). The difficulty often encountered when solving for $\mathbf{b}$ is that the $\mathbf{R}^T\mathbf{R}$ matrix is not invertible (see Equation 5.23) because of redundancy in the variables. Principal component regression eliminates this redundancy by constructing a new matrix **U** with columns that are linear combinations of the original columns in **R**. This matrix has the same number of rows as **R**, but fewer columns. Using the **U** matrix, a new model is written as shown in Equation 5.29.

$$\mathbf{c} = \mathbf{U}\tilde{\mathbf{b}} \qquad (5.29)$$

where $\mathbf{U}^{\mathrm{T}}\mathbf{U}$ is now invertible.

Various approaches can be taken for constructing the **U** matrix. With PCR, a principal components analysis is used because PCA is an efficient method for finding linear combinations of variables that describe variation in the row space of **R** (See Section 4.2.2). With analytical chemistry data, it is usually possible to describe the variation in **R** using significantly fewer PCs than the number of original variables. This small number of columns effectively eliminates the matrix inversion problem.

The **U** matrix is the score matrix from PCA, which defines the location of the samples relative to one another in row space. Therefore, **U** can be thought of as the output from an instrument which has the principal components as the measurements. The score matrix is related to the original matrix **R** in the following manner:

$$\mathbf{R} = \mathbf{U}\mathbf{S}\mathbf{V}^{\mathrm{T}} \qquad (5.30)$$

where **U** is the score matrix, **V** is a matrix containing the loadings, and **S** is a diagonal matrix of singular values (i.e., all off-diagonal elements are zero). The diagonal elements contain the information about how much variance each principal component describes. The orthonormal property of **V** (i.e., $\mathbf{V}^{\mathrm{T}}\mathbf{V} = \mathbf{I}$) can be used to solve Equation 5.30 for **U** as follows:

$$\mathbf{U} = \mathbf{R}\mathbf{V}\mathbf{S}^{-1} \qquad (5.31)$$

As a result of the principal component calculation, the **U** matrix has a number of columns equal to the minimum of the number of samples or variables. Knowing that only some of the columns in **U** contain the relevant information, a subset is selected. Choosing the relevant number of PCs to include in the model is one of the most important steps in the PCR process because it is the key to the stabilization of the inverse. Ordinarily the columns in **U** are chosen sequentially, from highest to lowest percent variance described.

One mathematical reason for reducing the number of columns in **U** can be seen upon examination of Equation 5.31. This equation shows that **U** is a function of $\mathbf{S}^{-1}$, which is calculated by taking the inverse of each of the diagonal elements of **S**. The calculation of $\mathbf{S}^{-1}$ is not stable if PCs with small diagonal elements (which likely describe only noise) are retained. This rank selection issue is discussed in much more depth in the examples below.

Given a truncated $\underset{\sim}{\mathbf{U}}$ matrix, it is now possible to solve Equation 5.29 for the regression vector $\underset{\sim}{\mathbf{b}}$:

$$\hat{\underset{\sim}{\mathbf{b}}} = \mathbf{U}^T\mathbf{c} \tag{5.32}$$

where the orthonormal property of $\mathbf{U}$ has been used and $\hat{\underset{\sim}{\mathbf{b}}}$ is the estimate of $\underset{\sim}{\mathbf{b}}$.

This regression vector can then be used to predict the concentration in an unknown sample using a two-step process. First, given the measurement of the unknown ($\mathbf{r}_{un}$) and the $\mathbf{V}$ and $\mathbf{S}$ from the calibration, the score vector for the unknown ($\mathbf{u}_{un}$) is obtained using Equation 5.31:

$$\mathbf{u}_{un} = \mathbf{r}_{un}\,\mathbf{VS}^{-1} \tag{5.33}$$

Second, this $\mathbf{u}_{un}$ and $\hat{\underset{\sim}{\mathbf{b}}}$ from Equation 5.32 are used to predict the concentration in the unknown sample using Equation 5.29:

$$\hat{c} = \mathbf{u}_{un}\hat{\underset{\sim}{\mathbf{b}}} \tag{5.34}$$

Summarizing to this point, PCR has been used to stabilize the regression by replacing the original measurement variables with a subset of the columns of the score matrix. The scores of the calibration spectra are estimated and the concentration vector is regressed on these scores to estimate regression coefficients. To perform prediction, the score vector of the unknown spectrum is calculated using the principal components estimated from the calibration samples. After the score vector is estimated, the regression coefficients are used to estimate the concentration of the unknown sample.

Although prediction is presented above as a two-step process, these can be combined into one step. If $\mathbf{u}_{un}$ in Equation 5.34 is replaced with the right side of Equation 5.33, the following one-step prediction equation is found:

$$\hat{c} = \mathbf{r}_{un}(\mathbf{VS}^{-1}\hat{\underset{\sim}{\mathbf{b}}}) \tag{5.35}$$

This equation shows that even though the PCR model is written to relate the score vectors to concentrations (Equation 5.29), it is a linear combination of all variables that are being used to estimate the concentrations (as seen in Equation 5.19).

Differences Between PLS and PCR: Principal component regression and partial least squares use different approaches for choosing the linear combinations of variables for the columns of $\mathbf{U}$. Specifically, PCR only uses the $\mathbf{R}$ matrix to determine the linear combinations of variables. The concentrations are used when the regression coefficients are estimated (see Equation 5.32), but not to estimate $\mathbf{U}$. A potential disadvantage with this approach is that variation in $\mathbf{R}$ that is not correlated with the concentrations of interest is used to construct $\mathbf{U}$. Sometimes the variance that is related to the concentrations is a very small portion of the overall variation. If this is the case, PCR may fail to find appropriate linear combinations of variables for modeling the concentrations.

With PLS, the covariance of the measurements with the concentrations is used in addition to the variance in **R** to generate **U**. This distinction between PLS and PCR is depicted in Figure 5.89 where the only difference between the methods is in the first step (estimation of **U**). As with PCR, PLS is not commonly implemented using a two-step process, but is presented this way for clarity. For more details, see Martens and Næs (1989).

Although the PCR and PLS algorithms are different, we do not feel there is overwhelming evidence to suggest that one method is superior to the other. Therefore, we do not have a strong recommendation when choosing between PCR and PLS. In practice, our tendency is to use PLS because of our experiences with it and the software tools we employ. Therefore, the examples discussed below only present the PLS results. [See Haaland and Thomas (1990) for additional readings.]

An additional point regarding PLS is that there are two approaches that can be used to develop a calibration model if there is more than one analyte of interest. Using the approach termed PLS1, separate models are developed for each analyte. With PLS2, a single model is constructed using all the analytes simultaneously. This second approach complicates the selection of rank for the final model, and therefore, PLS1 is the approach used in this book and the one we recommend. A similar distinction does not exist for PCR (i.e., there is no PCR1 and PCR2) because **U** is determined independent of the concentrations of the different analytes.

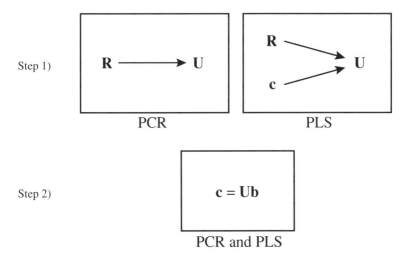

FIGURE 5.89. Illustration of the difference between PCR and PLS. The major difference between the methods is how the surrogate matrix (**U**) is calculated.

5.3.2.1 PLS EXAMPLE 1   In this project, a PLS model is constructed for the prediction of component A in a chemical process containing components A, B, and C using spectral measurements. The concentration of component B can be controlled on-line. However, the concentration of component C cannot be determined off-line and cannot be quantitatively manipulated by process settings. Given the limitations on the determination of component C, the use of an inverse model is the best approach for making a model to predict the concentration of A.

The concentration range for components A and B are both from 10 to 30 units and can be varied in the process for the calibration experiments. Because the level of C cannot be controlled, it is assumed that sufficient variation in the level of C is captured during the data collection. This is necessary in order to implicitly model the variation of C. However, if C is not effectively modeled, the prediction diagnostics will indicate the deficiency.

There are many things to consider when choosing an experimental design for the calibration data. See chapter 2, Appendix A for special considerations for inverse models. The design used for this calibration is a three-level, full-factorial design (for components A and B) with two repeats of the center point, giving a total of 11 calibration samples. The process settings are manipulated to prepare the calibration samples specified by the design. One or more spectra are measured at each design point. The reference concentration values for the calibration samples are then derived from the process settings.

After the calibration data are collected, PLS is used to construct the model relating spectra to the concentrations of component A. The discussion follows the "Six Habits of an Effective Chemometrican" which are detailed in Chapter 1.

Habit 1. Examine the Data
Both the concentration and measurement raw data should be examined. It is important to see if the concentrations obtained are near the concentrations specified in the design and if the measurements look reasonable. Figure 5.90 is a plot of the reference concentrations which match well with the three-level factorial design specified. The corresponding raw spectral data plotted in Figure 5.91 appear to be reasonable (i.e., no apparent anomalous samples or variables).

Habit 2. Preprocess as Needed
From an examination of the raw data, no preprocessing is indicated. (See Chapter 3 for a discussion of preprocessing.)

Habit 3. Estimate the Model
To build a calibration model, the software requires the concentration and spectral data, preprocessing options, the maximum rank (number of factors) to estimate, and the approach to use to choose the optimal number of factors to include in the model. This last option usually involves selection of the cross-validation technique or the use of a separate validation set. The maximum

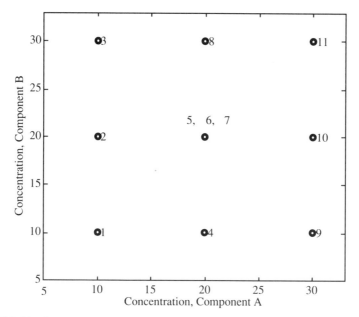

FIGURE 5.90. Concentration of component B versus concentration of component A in a three-level, full-factorial experimental design.

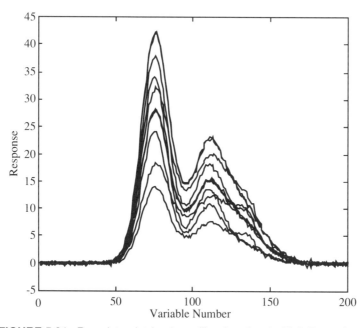

FIGURE 5.91. Raw data plot for the calibration data in PLS Example 1.

rank must be less than or equal to the minimum of the number of calibration samples or the number of variables. It is often selected to be smaller than this number to reduce calculation time. However, it is important to select a large enough maximum rank so that the optimum rank can be determined.

Habit 4. Examine the Results/Validate the Model

Several diagnostic tools are discussed below and a summary is found at the end of the section in Table 5.18. These tools are used to investigate three aspects of the data set: the model, the samples, and the variables. The headings for each tool indicate the aspects that are studied with that tool. The primary use of the model diagnostic tools is to determine the optimum rank of the model. The sample diagnostic tools are used to study the relationships between the samples and identify unusual samples. The variable diagnostic tools do the same, but for the variables.

Percent Variance Table (Model Diagnostic): The first diagnostic is the percent variance explained for both the concentration and the measurement data. The results for this example are listed in Table 5.15 as a function of the number of factors included in the model. The percent variance explained is the amount of variance explained by a model with a given number of factors relative to the total variance in the data set.

Some understanding of the size of the model that should be used (i.e., the number of factors to include in the model) can be gained by examining this table. Note that for PLS and PCR each successive factor does not always have to explain a decreasing amount of concentration variance. For example, in Table 5.15 the second factor explains 26.33% of the concentration variation while the first factor explains only 25.76%. For the measurement data ($\mathbf{R}$), the percent variance explained per factor for PCR always decreases; this is a property of PCA (see Section 4.2.2). For PLS, a strict decrease in the percent variance explained for $\mathbf{R}$ is not always observed.

TABLE 5.15. Percent Variance Explained for Component A PLS Model

Factor No.	Percent Concentration Variance Explained (C Matrix)		Percent Measurement Variance Explained (R Matrix)	
	Each Factor	Cumulative	Each Factor	Cumulative
1	25.76	25.76	87.80	87.80
2	26.33	52.10	11.71	99.51
3	22.87	74.97	0.18	99.69
4	24.55	99.52	0.07	99.77
5	0.45	99.97	0.04	99.81
6	0.03	100.00	0.03	99.84
7	0.00	100.00	0.04	99.88

For this example, much of the spectral variance has been explained after two factors (99.51%) with only 52.1% of the concentration variance explained. The fact that only a small amount of the concentration variance is accounted for may indicate a problem with the data, such as significant error in the concentration reference values or the lack of a Beer's law relationship between concentrations and spectral responses.

The percent variance explained can also be used to help determine the number of factors to use in the model by comparing the percent variance remaining with a known noise level in the spectral and/or concentration data.

Root Mean Square Error of Prediction (RMSEP) Plot (Model Diagnostic): Prediction error is a useful metric for selecting the optimum number of factors to include in the model. This is because the models are most often used to predict the concentrations in future unknown samples. There are two approaches for generating a validation set for estimating the prediction error: internal validation (i.e., cross-validation with the calibration data), or external validation (i.e., perform prediction on a separate validation set). Samples are usually at a premium, and so we most often use a cross-validation approach.

The statistic used to quantify the error in prediction is the root mean square error of prediction (RMSEP):

$$\text{RMSEP} = \sqrt{\frac{\sum_{i=1}^{nsamp} (c_i - \hat{c}_i)^2}{nsamp}} \qquad (5.36)$$

where c_i is the concentration of the ith prediction sample, $\hat{c}_i$ is the predicted concentration for this sample, and nsamp is the number of prediction samples.

When using a cross-validation approach, the original calibration data set is divided into calibration and prediction subsets. (Wold, 1978; Stone, 1974; Eastment and Krzanowski, 1982) For example, the data set can be split in half using the first half of the data to construct a calibration model for predicting the second half. The roles of the data subsets are then exchanged and a model is constructed where the second half is used to predict the first half. In this way, each of the samples in the data set is predicted once. Another common approach is called leave-one-out cross-validation, where the data set is divided into subsets containing nsamp-1 and one sample. The nsamp-1 samples are used to build the model and the one left out sample is used for prediction. This process is repeated until each sample is left out once. For large data sets, this can be computer intensive and, therefore, it can be more practical to leave out larger subsets. Keep in mind that there is a limit to how many and which samples may be left out because the calibration set must contain an adequate set of samples to construct the model at each step.

Another situation where care must be taken when splitting the data set is when replicate samples are present in the design. For example, if there are three replicates at each design point and two of the three replicates are re-tained for the calibration set, the prediction of the third replicate can yield an overly optimistic assessment of model performance. This is because real pre-diction samples will not typically fall on a calibration design point. On the other hand, when entire design points are left out, the model will extrapolate for the extreme points in the design, which leads to a conservative assessment of model performance. Therefore, it is best to use several different leave-out strategies to obtain the most accurate assessment of predictive ability. (Cruciani et. al., 1992)

To select an optimum rank, the RMSEP is evaluated with models con-structed using different numbers of factors. The RMSEP decreases when the predicted value $(\hat{c}_i)$ is close to the known value (c_i) and, therefore, small RMSEP values are desired (see Equation 5.36). To choose the optimum num-ber of factors, a plot of RMSEP versus the number of factors is examined. This plot typically presents results from models constructed using more factors than are expected to be significant.

The idealized behavior for the RMSEP plot is shown in Figure 5.92a. As the complexity of the model increases (i.e., the number of factors increases), the signal relevant for prediction of the component of interest (including interfer-ent signal that must be modeled) is incorporated into the model. This results in a decrease in the prediction error. As more factors are used, more noise is also incorporated into the model. This increases the statistical uncertainty in the model, which leads to an increase in the prediction error. Choosing the correct number of factors is therefore a trade-off between incorporating rele-vant signal and excluding noise. For RMSEP plots like those shown in Figure 5.92a, too much noise is added when the addition of a factor results in an in-crease in the RMSEP. With well-behaved RMSEP plots, we recommend choos-ing the number of factors that corresponds to the first minimum in the RMSEP plot as a starting point. This choice is then refined using the other diagnostics discussed below.

Figure 5.92b displays a more commonly observed form of the RMSEP plot where RMSEP continually decreases as factors are added. This implies that one should use all of the calculated factors to construct the model (or calculate even more). This is not the correct answer even though the RMSEP continues to decrease because the change is usually not statistically significant. The key is to determine where the plot is "leveling off" and select the number of fac-tors at this point. Statistical tests can be used to determine when the decrease in RMSEP is not significant (Haaland and Thomas, 1988; Osten, 1988; van der Voet, 1994). We recommend using these statistics in conjunction with the other diagnostics discussed below.

Figure 5.92c shows an RMSEP plot that displays erratic behavior. This type of plot is observed when the algorithm is not able to model the concentration variations. It can also result when gross errors are present in the reference val-ues (e.g., transcription errors in the concentration values, mixed up samples, and/or poor reference methods).

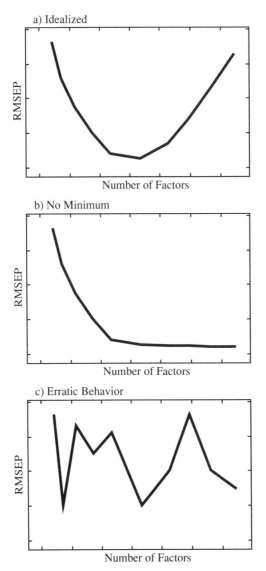

FIGURE 5.92. Possible behavior of RMSEP versus number of factors included in the model. *(a)* Idealized behavior; *(b)* behavior more typically observed; *(c)* behavior indicating problems with the data and/or model.

The leave-one-out cross-validation RMSEP plot for this example (shown in Figure 5.93) shows a clear minimum at two factors. The shape is not ideal because the RMSEP decreases again when the fourth factor is added.

The RMSEP is in the same units as the concentration reference values (see Equation 5.36). If the model performs well, the magnitude of the RMSEP

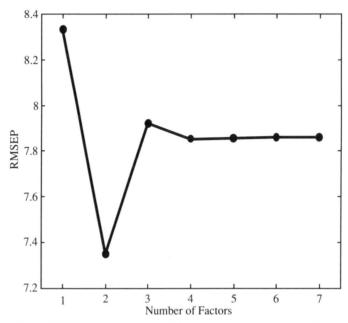

FIGURE 5.93. RMSEP versus number of factors for component A using leave-one-out cross-validation.

should be on the order of the error in the reference method (but not smaller). In this example, the RMSEP is indicating an expected error (1σ) in prediction of future samples on the order of 7.4 units using a two-factor model. This error is unacceptably large given the known reference error, which is 0.002, and the fact that the calibration range is 10–30.

A two-factor model is tentatively considered, but there is concern over the magnitude of the RMSEP. In practice, several models with different numbers of factors can simultaneously be considered. The choice of the number of factors to use for the final model is then refined by examining the different diagnostics that follow.

Predicted vs. Known Concentration Plot (Model and Sample Diagnostic): Figure 5.94 shows the cross-validation predicted concentrations for component A using a two-factor model. For model diagnostics, this plot is examined for bias or lack of fit. The values are expected to fall randomly about a line of slope one and zero intercept. See also Section 5.2.1.1 for more discussion of this diagnostic. Ideal behavior is not observed, which is in agreement with the large RMSEP.

This plot is also used to look for unusual samples. This is indicated if a small number of samples are significantly off of the ideal line. It is not possible to identify outliers because the majority of the samples do not fit the ideal line in this example.

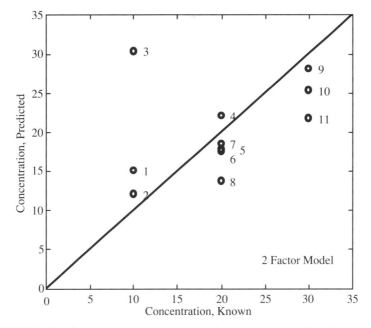

FIGURE 5.94. Predicted concentration versus known concentration for component A using leave-one-out cross-validation.

Concentration Residual vs. Predicted Concentration Plot (Model and Sample Diagnostic): Examination of the concentration residuals versus predicted concentration plot provides a different perspective of the prediction performance, as shown in Figure 5.95. Ideal residuals fall randomly about the line with zero slope and zero intercept. See also Section 5.2.1.1 for more discussion of this diagnostic. In Figure 5.95, sample number 3 is clearly predicted more poorly than the other samples. However, there is not enough information to determine the root cause of the problem. Furthermore, it is not advisable to assign root causes when the validity of the model is in question.

Concentration Residual vs. Other Postulated Variable Plot (Model and Sample Diagnostic): In some situations it is instructive to plot the concentration residuals versus other parameters such as run order or the levels of one of the other components. This can help identify the source of unmodeled variation. For this example, no variables were identified that correlate with the concentration residual.

Loadings Plot (Model and Sample Diagnostic): The loadings can be used to help determine the optimal number of factors to consider for the model. For spectroscopic and chromatographic data, the point at which the loading displays random behavior can indicate the maximum number to consider. Numerical evaluation of the randomness of the loadings has been proposed as a method for determination of the rank of a data matrix for spectroscopic data

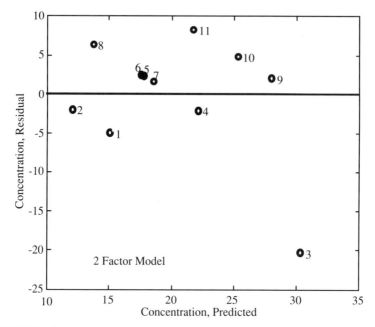

FIGURE 5.95. Concentration residual (known–predicted) versus predicted concentration for component A using leave-one-our cross-validation.

(Cattell, 1958). For this example, the first four loading vectors are plotted in Figure 5.96 (the offset was added for clarity). These loadings reveal some structure in the third and fourth loadings, although in general this plot is not inconsistent with the RMSEP plot which indicates a two-factor model.

The loadings can also be useful as a variable diagnostic. They represent the relative contribution of the variables in the model and may be expected to exhibit behavior similar in frequency to the original data. It is appropriate to investigate variables if the corresponding features in the loadings are of a very different frequency content than the original data. No unusual behavior is observed for the loadings in Figure 5.96, indicating no outlier variables.

Studentized Concentration Residual vs. Sample Leverage Plot (Sample Diagnostic): This diagnostic is used to more closely examine individual samples. The sample leverage for the ith sample is computed by

$$h_i = \frac{1}{\text{nsamp}} + \mathbf{u}_i^{\mathrm{T}}(\mathbf{U}^{\mathrm{T}}\mathbf{U})\mathbf{u}_i \qquad (5.37)$$

where nsamp is the number of samples in the calibration set, $\mathbf{u}_i$ is the score vector for sample i, and $\mathbf{U}$ is the score matrix from the calibration set truncated to the appropriate number of factors.

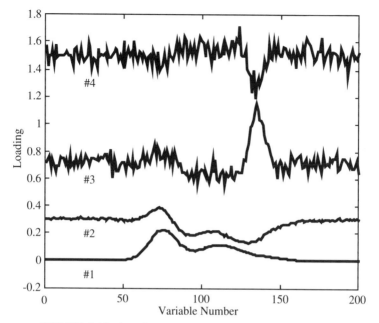

FIGURE 5.96. Loadings for the component A PLS model.

Sample leverage is a measure of how different the measurements of the ith sample are from the other samples in the data set. When calculated for a calibration data set, it has a maximum value of 1. Samples with leverage values that are significantly larger than the rest of the samples (i.e., greater than three times the average leverage) should be inspected more closely. Although they are not necessarily "bad" or "outliers," they do have a large influence on the model. These samples are investigated because an outlier sample that also has high leverage has a large adverse effect on the model. Note, however, that samples with high concentrations will also have high leverage, and these samples do not necessarily have an adverse influence on the model.

The other axis of the diagnostic plot is the studentized concentration residuals. These residuals are similar to the concentration residuals in Figure 5.95, but have been converted to standard deviation units as shown in Equation 5.38:

$$t_i = \frac{c_i - \hat{c}_i}{s\sqrt{1 - b_i}} \tag{5.38}$$

where t_i is the studentized concentration residual for the ith sample, c_i is the known concentration, $\hat{c}_i$ is the predicted concentration, b_i is the sample leverage calculated using Equation 5.37, and s is the standard deviation of the concentration residuals, computed as shown in Equation 5.39 where nfact is the number of PLS factors used.

$$s = \sqrt{\frac{\sum_{i=1}^{nsamp} (c_i - \hat{c}_i)^2}{nsamp - nfact - 1}} \tag{5.39}$$

The studentized residual measures how well the concentration of the ith sample is fit by the model. Large positive or negative studentized residual values indicate samples that are not fit well. In the following discussion, studentized residuals greater than ± 2.5 are considered to be large. This is because the studentized residuals are in units of standard deviations from the mean value, and ± 2.5 is unusual given standard statistical assumptions.

Some software packages include on the plot of concentration residual versus leverage a vertical line indicating the average sample leverage, a horizontal line at zero studentized concentration residual, and a pair of horizontal lines associated with a specified confidence level for the studentized concentration residual. The interested reader is referred to the literature for more details on hypothesis testing for outliers using studentized residuals (Weisberg, 1985).

Table 5.16 displays the four possibilities for a sample in the studentized concentration residuals versus sample leverage plots. A typical sample will have small sample leverage (leverage <3*average calibration leverage) and small studentized concentration residual (i.e., | studentized residual | <2.5). These samples have measurements that are not out of the ordinary compared with the rest of the calibration samples and concentrations that are predicted well. A small sample leverage and large studentized residual indicates a measurement vector which is similar to those in the calibration set but one that yields a poor concentration prediction. Because the small leverage does not indicate a problem with the measurement vector, this combination often indicates a problem with the reference concentration values.

TABLE 5.16. Possible Combinations for Sample Leverage and Studentized Concentration Residuals

	Small Leverage Large Studentized Residual Concentration Error ?	Large Leverage Large Studentized Residual C and/or R Error ?
Studentized Residual	Small Leverage Small Studentized Residual Sample OK	Large Leverage Small Studentized Residual Spectral Error? Extreme Concentration?

Sample Leverage

A large sample leverage and small studentized concentration residual indicates a measurement vector that is different from the rest of the data, yet the model is predicting the concentration well. This may or may not indicate a problem with this particular sample. Remember that a sample is not necessarily an outlier just because it has a large leverage. It may simply be at the limits of the design space and thus expected to have large leverage. On the other hand, the reason the sample is predicted well may be because the large leverage has forced the model to fit this sample well. (The leverage and studentized concentration residuals are based on using all of the calibration data, or model fit, not on cross-validation).

The last possible combination is to have a sample with large sample leverage and large studentized concentration residuals. This indicates a problem with the measurement vector and/or the reference concentration values. Because of the possibility of the model being forced to fit influential points, the recommendation is to reexamine the measurement and reference concentration values whenever a sample has large leverage.

For this example, Figure 5.97 is the plot of studentized residual versus sample leverage for the calibration samples using a two-factor PLS model. Sample 3 has a reasonable leverage and a large studentized concentration residual with an absolute value near 3. Given that there are no apparent problems with the measurement vector, the reference value is suspect. Further supporting this hypothesis, the studentized residual versus sample leverage plot for

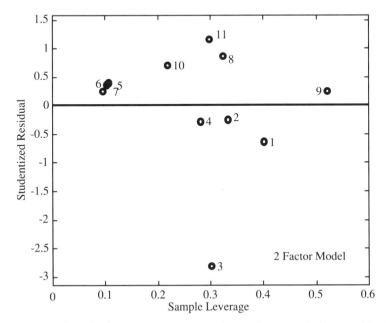

FIGURE 5.97. Sample leverage versus studentized concentration residuals for component A PLS model.

component B (not shown) does not indicate a problem with sample 3. Errors in the measurement vector would have been manifested in both of these plots as a large leverage.

All other samples have reasonable leverages and acceptable studentized residuals.

Scores Plot (Sample Diagnostic): The score plots show the relationship of the samples in PLS row space and are examined for consistency with what is known about the data set. Look for unusual or inconsistent patterns which can indicate potential problems with the model and/or samples (see also PCA, Section 4.2.2). In the PCA discussion the scores are referred to as PCs, but in PLS they are referred to as factors.

Figure 5.98 shows the Factor_2 versus Factor_1 plot for the example data set which describes 99.5% of the variance in the measurements. The experimental design can be seen in this plot (compare Figure 5.98 to Figure 5.90). The center-point samples (5,6,7) are close to one another and the slight difference in position may be due to measurement errors or differing amounts of the third component (which is being implicitly modeled). Samples 3 and 11 are on top of each other in the far right side of the plot and there is a hole in the lower left corner of the design. This warrants further investigation because a full design was used and replicates were only run at the midpoint of the design. Previous diagnostics also indicated unusual behavior with sample 3. The fact that samples 3 and 11 are close to each other implies that they have similar spectral features. However, the reported concentrations of component A for samples 3 and 11 are 10 and 30, respectively (see Figure 5.90).

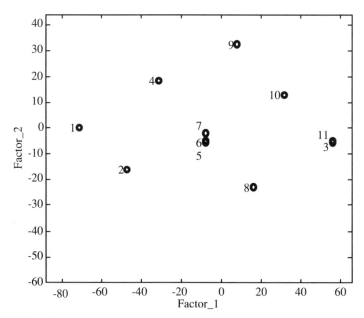

FIGURE 5.98. Factor_2 versus Factor_1 PLS scores plot for component A.

One way to resolve these inconsistencies is to plot the spectra of samples 3 and 11 and compare the differences in the raw data. Figure 5.99*a* displays the two spectra and the difference spectrum. The two spectra are the same except for slight differences which can be attributed to measurement noise. Sample 3 is known to be the problem given the known concentrations and the spatial relationship between the other samples in the scores plots (i.e., it should be in the lower center portion of the graph). This means that either sample 3 was incorrectly prepared to have the same concentration as sample 11, or sample 11 was measured when sample 3 was thought to have been measured. When the spectrum of sample 3 is remeasured, the resulting spectrum is very different from sample 11 (see Figure 5.99*b*).

Now that this problem has been identified and corrected, a new model is developed. The results for Habits 1–3 are identical except for the new sample 3 spectrum and, therefore, the following discussion begins with Habit 4.

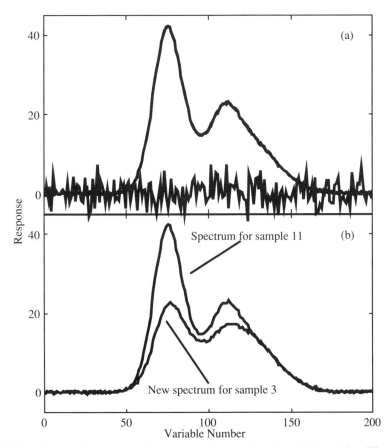

FIGURE 5.99. (*a*) Overlay plot of spectrum for samples 3 and 11 and the difference between the two spectra multiplied by 10. (*b*) New spectrum for sample 3 and original sample 11 spectrum.

TABLE 5.17. Percent Variance Explained for Component A, PLS Model Corrected Data

Factor No.	Percent Concentration Variance Explained (C Matrix)		Percent Spectral Variance Explained (R Matrix)	
	Each Factor	Cumulative	Each Factor	Cumulative
1	72.40	72.40	79.58	79.58
2	27.34	99.73	19.77	99.35
3	0.26	99.99	0.32	99.68
4	0.01	100.00	0.05	99.73
5	0.00	100.00	0.04	99.77
6	0.00	100.00	0.04	99.82
7	0.00	100.00	0.04	99.86

Percent Variance Table (Model Diagnostic): The percent variance explained for the corrected data is shown in Table 5.17. Comparing these results to those found in Table 5.15, a more continuous behavior of the percent variance explained is seen for the concentration data as a function of factor number. With three factors, almost all of the variance in the concentration data has been explained and there is just a small amount of spectral variance remaining. The results are reasonable given the fact that three chemical components are known to be varying in the samples.

Root Mean Square Error of Prediction (RMSEP) Plot (Model Diagnostic): The new RMSEP plot in Figure 5.100 is more well behaved than the plot shown in Figure 5.93 (with the incorrect spectrum 3). A minimum is found at 3 factors with a corresponding RMSEP that is almost two orders of magnitude smaller than the minimum in Figure 5.93. The new RMSEP plot shows fairly ideal behavior with a sharp decrease in RMSEP as factors are added and then a slight increase when more than three factors are included.

Predicted vs. Known Concentration Plot (Model and Sample Diagnostic): Figure 5.101 displays the predicted versus the known concentration for component A. This can be compared with the results shown in Figure 5.94 for the analysis of the faulty data set. Here the results lie near the ideal line (i.e., no evidence of problems with the model or individual samples are indicated).

Concentration Residual vs. Predicted Concentration Plot (Model and Sample Diagnostic): The concentration residuals displayed in Figure 5.102 are as expected (i.e., randomly arranged about zero residual). No obvious problems with any of the samples are indicated and the magnitude of the concentration residuals are much smaller than previously observed.

Concentration Residual vs. Other Postulated Variable Plot (Model and Sample Diagnostic): This plot is not included because no variables were found to correlate with the concentration residuals.

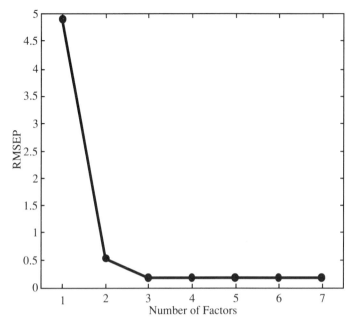

FIGURE 5.100. RMSEP versus number of factors for component A, corrected data, using leave-one-out cross-validation.

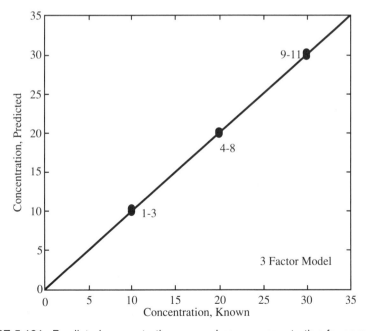

FIGURE 5.101. Predicted concentration versus known concentration for component A, corrected data, using leave-one-out cross-validation.

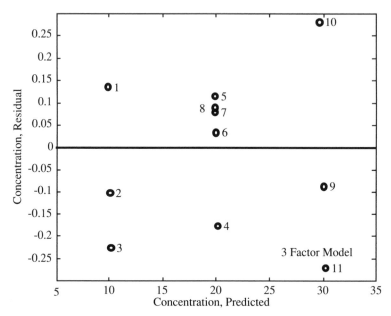

FIGURE 5.102. Concentration residuals versus the predicted concentration for component A, corrected data, using leave-one-out cross-validation.

Loadings Plot (Model and Sample Diagnostic): Figure 5.103 displays the first four loading vectors versus variable number for this data set (the offset was added for clarity). Loading vectors 1–3 are reasonable given the features observed in the raw data plot. Loading 3 has larger noise than the other two, but is apparently correcting for a feature that is present at variable number ~130. The fourth loading vector resembles random noise, which supports the choice of three factors for the model. This plot does not indicate any problems with the model or any individual variables. Be cautious about extensive interpretation of loading vectors. They are abstract mathematical representations of the data space confined to be orthogonal and, therefore, often do not have readily interpretable features.

Studentized Concentration Residual vs. Sample Leverage Plot (Sample Diagnostic): Figure 5.104 displays the studentized concentration residuals versus the sample leverage for the calibration samples. There are no samples with extremely large studentized residuals. The samples with the largest leverage values (1, 3, 6, 9, and 11) correspond to the samples at the extreme corners of the design and one center point. Sample number 6 has a higher leverage than is expected from a center-point sample. This is probably due to a large amount of component C. The conclusion from this plot is that the model is working well and none of the samples appear to be unusual.

Scores Plot (Sample Diagnostic): Figure 5.105 displays Factor_2 versus Factor_1, describing 99.4% of the spectral variance for the corrected data.

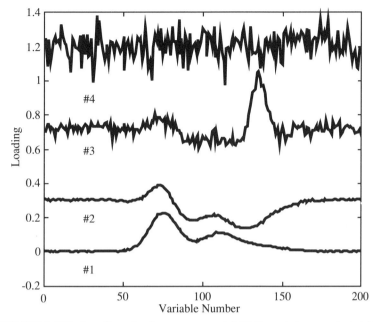

FIGURE 5.103. Loadings for the component A PLS model, corrected data.

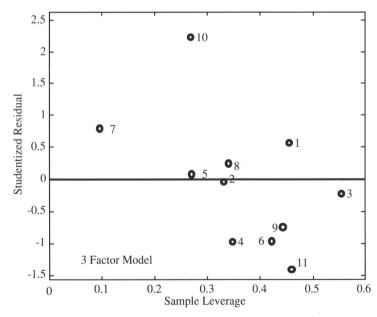

FIGURE 5.104. Studentized residuals versus sample leverage for three-factor PLS model for component A, corrected data.

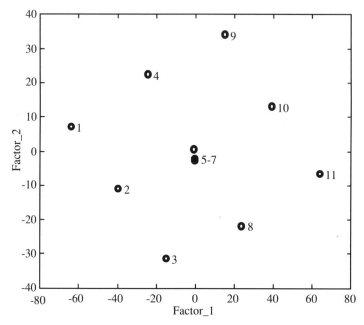

FIGURE 5.105. PLS factor_2 versus factor_1 for component A, corrected data.

The complete design is seen in the score space with replicate center points clearly visible. Note that the interpretation of scores plots is not always as straightforward as in this example. The experimental design is not seen if the experiment is not well designed or if the problem is high dimensional. The level of implicitly modeled components (e.g., component C) also has an effect on the relative position of the samples in score space. For this example, the effect of C on the relative placement of the samples in score space is small.

Measurement Residual Plot (Sample and Variable Diagnostic): Figure 5.106 displays the measurement residuals for the three-factor PLS model. The residuals are that portion of the sample measurement vector that is not explained using a given number of factors. The residuals in Figure 5.106 are quite random and the magnitude is small relative to the original measurements (maximum response ~40). These residuals are random, but it is fairly common to observe structured residuals with an inverse model such as PLS. When constructing the calibration models, these methods can ignore variation that is not correlated with the concentration values of the component of interest. This leads to residual spectra with structure. If structure is observed, determine if it is for only a few or nearly all of the samples. If only a few samples have large and/or structured residuals, this may indicate a problem with those samples.

Residual spectra can also be used to identify variables that appear to be anomalous. If there are certain variables with large residuals compared to the other variables, these variables may be problematic.

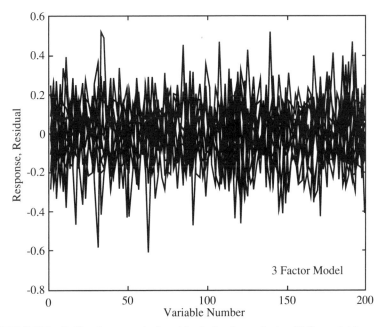

FIGURE 5.106. Calibration spectral residuals for three-factor PLS model for component A, corrected data.

F_{calc} *Plot (Sample Diagnostic):* A convenient comparison of the spectral residuals from sample to sample is found by summarizing them into a single metric called F_{calc}. First a sum of squares value is calculated for the validation samples:

$$ss_{cal} = \sum_{i=1}^{nsamp} \sum_{j=1}^{nvars} (r_{ij} - \hat{r}_{ij})^2 \qquad (5.40)$$

where $(r_{ij} - \hat{r}_{ij})$ is the cross-validation spectral residual value for the ith calibration sample and jth variable. This sum of squares reflects the magnitude of all of the residual spectra.

A sums of squares is then calculated for each individual sample. The residual of a sample is a vector, and the sum of squares summarizes it in a single number, as shown in Equation 5.41:

$$ss_i = \sum_{j=1}^{nvars} (r_{ij} - \hat{r}_{ij})^2 \qquad (5.41)$$

To form an F value (labeled F_{calc}), the two sums of squares are divided by the appropriate degrees of freedom and ratioed as shown in Equation 5.42:

$$F_{calc} = \frac{ss_i}{df_i} \bigg/ \frac{ss_{cal}}{df_{cal}} \tag{5.42}$$

Owing to the highly correlated nature of many analytical measurement systems, the appropriate number of degrees of freedom to use is debatable and difficult to gauge. One approach is to define the degrees of freedom as df_i = nvars − nfact and df_{cal} = nsamp∗nvars − nvars − nfact∗max[nsamp,nvars] (Martens and Jensen, 1982). It has also been suggested that the exact degrees of freedom estimate is not critical for satisfactory outlier detection (Martens and Næs, 1989). We use a simple approach, df_i = 1 and df_{cal} = nsamp, as recommended by other researchers (Haaland and Thomas, 1988).

The statistical approach is to compare F_{calc} to a statistically determined F_{crit} and question any sample where F_{calc} exceeds F_{crit}. We take a more heuristic approach that does not rely on debatable degrees of freedom. We compute the average and standard deviation of the F_{calc} values from the calibration and formulate an F_{crit} as being the average F_{calc} value plus three times the standard deviation. The software package in use may employ another approach for determining F_{crit}. This critical value is used to evaluate if there are outliers in the calibration and validation data sets. It is also used in the prediction phase to automate the identification of unknown samples with unusually large spectral residuals (see Habit 6).

Figure 5.107 shows the calibration F_{calc} values determined using leave-one-out cross-validation. The F_{crit} value of 1.03 is calculated as the mean of the F_{calc} values plus three times the standard deviation of the F_{calc} values. There are no samples that are outside the critical range. The calculated critical value appears to be reasonable when compared to the distribution of the F_{calc} values.

Summary of Validation Diagnostic Tools for PLS/PCR, Example 1: Eleven samples were used to build a PLS model to predict the concentration of component A with varying amounts of components B and C. The details of the data set used to construct the calibration model are as follows:

Calibration design for components A and B: three-level full factorial

Calibration design for component C: natural

Validation design: leave-one-out cross-validation

Preprocessing: none

Variable range: 200 measurement variables

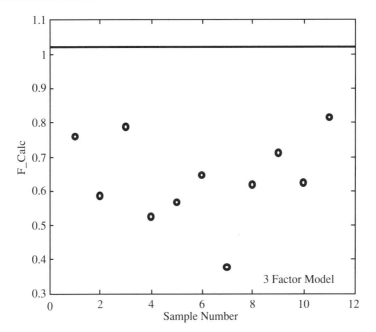

FIGURE 5.107. F_{calc} versus sample number for three-component PLS for component A, corrected data. The horizontal line is F_{crit}.

The operating ranges of the prediction model based on the calibration data set follow. Predicting future samples from outside this operating range is extrapolating and may produce unreliable results.

Components A and B: 10–30 units

Component C: unknown (allowed to vary naturally)

A three-factor PLS model was found to be optimal based on the model diagnostic tools. The measures of performance for the final model are as follows (see Table 5.18 for a description of these figures of merit):

RMSEP = 0.16 units

$F_{crit} = 1.03$

Average leverage = 0.36

Expected error in prediction = ±0.25 units

(Based on errors in predicted A concentration. See Figure 5.102.)

It was found through examination of the diagnostics that the spectrum of sample 3 was incorrectly measured. After the sample was remeasured, a rank

TABLE 5.18. Summary of Validation Diagnostics for PLS/PCR

Diagnostics	Description and Use
Model	
Percent variance table (% variance explained vs. number of factors)	Fraction of the total variation explained for given numbers of factors. Calculated for both the concentrations and the measurement vectors.
	Helps to determine the number of relevant factors (rank).
	With well-behaved data and an appropriate number of factors, the percent of the concentration variance not described is comparable to the known errors in the concentrations.
	A problem is indicated if most of measurement variance is explained and a considerable amount of concentration variance remains unexplained.
Root mean square error of prediction (RMSEP) plot (RMSEP vs. number of factors)	RMSEP is a single value that quantifies the magnitude of the concentration residuals for the validation samples for a given number of factors included in the model.
	Is used to help determine the optimal number of factors to include in the model.
	The ideal RMSEP plot decreases as factors are added into the model and increases when it reaches a value that is comparable to the known concentration errors.
	The optimum number of factors to include is the minimum or the point at which the RMSEP levels off.
Predicted vs. known concentration plot ($\hat{c}$ vs. c)	Predicted concentrations of validation samples plotted versus the known values.
	Used to determine whether the model accounts for the concentration variation in the validation set.
	Points are expected to fall on a straight line with a slope of one and a zero intercept.
	A modeling problem is indicated when a systematic pattern away from the ideal line is observed.
Concentration residual vs. predicted concentration plot $[(c - \hat{c})$ vs. $\hat{c}]$	The differences between the known and predicted concentrations (residuals) are plotted versus the predicted concentrations for validation samples.
	Used to determine whether the model accounts for the concentration variation in the validation set.
	Points are expected to vary randomly about a line of zero slope and zero intercept.
	A modeling problem is indicated when a systematic pattern away from the ideal line is observed.
	Patterns are often more discernible in this plot compared to the predicted versus known plot.
Concentration residual vs. other postulated variable plot $[(c - \hat{c})$ vs. $x]$	The concentration residuals can be plotted versus a postulated variable such as run order.
	Used to determine if the model is accounting for the variability correlated with the postulated variable.
	Points are expected to vary randomly about a line of zero slope and zero intercept.

TABLE 5.18. *(Continued)*

Diagnostics	Description and Use
	A modeling problem is indicated when a systematic pattern away from the ideal line is observed.
	Identified effects should be controlled or added to the model if possible.
Loadings plot (loadings vs. variable number)	For a given factor, the absolute magnitude of the loading for a measurement variable defines how that variable contributes to the factor. The patterns within the loading vectors from different factors are examined.
	When the measurement vectors are continuous in nature (e.g., infrared spectra) the loadings can help determine the number of relevant factors.
	A factor corresponding to a loading vector that resembles random variation is not considered relevant.
Sample	
Predicted vs. known concentration plot ($\hat{c}$ vs. c)	Predicted concentrations of validation samples plotted versus the known values.
	This plot can be used to identify outlier samples.
	Points are expected to fall on a straight line with a slope of one and a zero intercept.
	An outlier is indicated when a point is far from the ideal line relative to the other points.
Concentration residual vs. predicted concentration plot [$(c - \hat{c})$ vs. $\hat{c}$]	The differences between the known and predicted concentrations (residuals) are plotted versus the predicted concentrations for the validation samples.
	This plot can be used to identify outlier samples.
	Points are expected to vary randomly about a line of zero slope and zero intercept.
	A small number of samples with unusual concentration residuals is an indication of sample rather than model problems.
	Outliers are often more discernible in this plot compared to the predicted versus known plot.
Studentized concentration residual vs. sample leverage plot	Studentized concentration residuals are concentration residuals that have been divided by the concentration standard error of estimate and $\sqrt{1 - \text{leverage}}$. Sample leverage is a measure of the influence a sample measurement vector has on the model.
	Used to identify samples that warrant further investigation. (See Table 5.16 for a more detailed discussion on interpretation.)
Scores plot [Factor_y vs. Factor_x (vs. Factor_z)]	The scores represent how the samples are related to each other given the measurement variables and the PCR/PLS model. The plots are two- or three-dimensional representations of the row space.
	Used to examine the differences and similarities between the samples in the data set.

continued

TABLE 5.18. *(Continued)*

Diagnostics	Description and Use
	Closeness of samples in the plot is interpreted as chemical similarity. The location of the samples is expected to be consistent with any a priori information.
	Be sure to note the percent variance described by each factor when making interpretations.
Measurement residual plot [$(\mathbf{r} - \hat{\mathbf{r}})$ vs. variable number]	The measurement residuals are the portion of the sample measurement vector that is not explained using a given number of factors.
	Used to identify outlying samples.
	An outlier is identified as a sample with a residual vector that is significantly different from the other samples.
	Examining the variables where a sample has unusual features can sometimes be used to determine the cause of the problem.
	When using an inverse model such as PLS, structured measurement residuals are not uncommon.
F_{calc} plot (F_{calc} vs. sample number)	F_{calc} is a quantitative measure of the reliability of the concentration prediction of a validation sample. It is based on the measurement residual.
	It is used to identify samples with unusual measurement vectors.
	An unusual sample is indicated when the F_{calc} value is large relative to the F_{calc} values for the other validation samples.
	A critical value (F_{crit}) is established during model validation. This critical value is used to evaluate the reliability of future predictions.
Variable	
Loadings plot (loadings vs. variable number)	For a given factor, the absolute magnitude of the loading for a measurement variable defines how that variable contributes to the factor.
	Used to identify both important and unusual variables.
	A variable that contributes much to a factor has a large absolute loading (max = 1). Because the structure of the loading vector depends on the data being analyzed, the definition of an unusual variable is problem dependent.
	For well-behaved continuous data, the features of the loadings vectors are comparable to the preprocessed data.
Measurement residual plot [$(\mathbf{r} - \hat{\mathbf{r}})$ vs. variable number]	The residuals are the portion of the sample measurement vector that is not explained using a given number of factors.
	Used to identify unusual variables.
	Problematic variables are identified as those with unusual residual features for all of the samples.

three model was shown to have acceptable model performance based on the requirements of the application. No additional outliers were detected and the model implicitly accounts for variations due to components B and C.

Table 5.18 summarizes the validation diagnostic tools discussed in this section. The first column in the table lists the name of each tool and the second column describes results from both well-behaved and problematic data.

Habit 5. Use the Model for Prediction

Prediction is a simple vector multiplication of the regression vector by the preprocessed spectrum of the unknown to yield a concentration estimate (see Equation 5.25). Using this procedure, the predicted values for component A are obtained for 20 unknown samples.

Habit 6. Validate the Prediction

Several prediction diagnostic tools are discussed below and a summary is found at the end of the section in Table 5.19. These tools are similar to those found in the model-validation phase with the omission of any diagnostic that relies on knowledge of the true concentration values.

Predicted Concentration Plot: Figure 5.108 displays the predicted calibration concentration values for component A along with horizontal lines which represent the calibration concentration range for component A. All of the prediction samples fall within the calibration range for component A except for

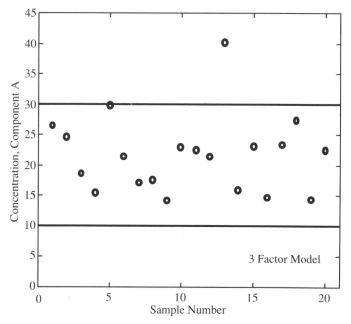

FIGURE 5.108. Prediction of component A in the unknowns. The horizontal lines indicate the range of concentrations in the calibration set.

sample 13. The prediction for this sample may not be reliable because the model is extrapolating.

Sample Leverage Plot: This prediction diagnostic replaces the studentized residual versus sample leverage plot because the studentized residual for an unknown sample cannot be calculated (the true concentration is not known). A large leverage for an unknown sample means the score values are unusual relative to the calibration samples. One difference between calibration and prediction leverage is that while the former has a maximum value of 1, the latter can be greater than 1.

For this example, the prediction sample leverages are shown in Figure 5.109 with a horizontal line denoting three times the average leverage of the calibration samples. In this case, all of the samples are below the calibration maximum and are comparable with each other except for sample 13. This is another indication that this sample is outside the calibration range.

F_{calc} *Values:* Whereas the sample leverage is based on the score vector of a sample, the F_{calc} value is based on the residual spectrum. The F_{calc} values are a convenient screening diagnostic for identifying samples that need closer examination.

Figure 5.110 displays the F_{calc} versus sample number for the 20 prediction samples. Three samples (1, 10, and 18) have F_{calc} values that are well above the F_{crit} value of 1.03 computed during the calibration phase. It may be confusing that these three samples have small leverage values but large F_{calc} values.

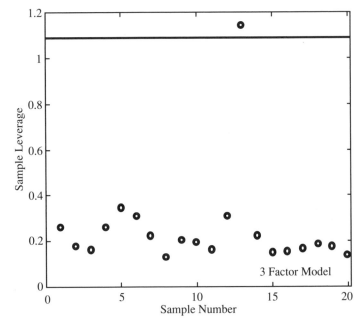

FIGURE 5.109. Sample leverage for the prediction data. The horizontal line represents three times the average leverage from calibration.

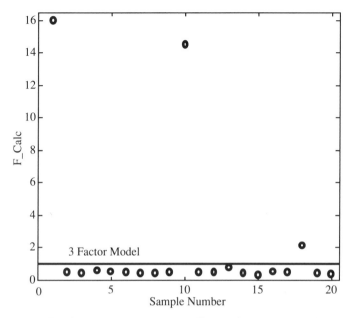

FIGURE 5.110. F_{calc} for the prediction data. The horizontal line is F_{crit} determined from calibration.

This occurs because the score vector is used to calculate the leverage and is also the portion of a spectrum that is used by the model for prediction. Therefore, the reasonable leverages for these three unusual samples indicate that the portion of these spectra that is used by the PLS model is reasonably similar to the spectra of the calibration samples. On the other hand, the residual spectrum (and the corresponding F_{calc} value) is the part of the spectrum that is not used by the model. The conclusion is that there is significant variation in these prediction spectra that is different than the calibration samples. However, the model is extracting a portion of the spectra that appears to be reasonable. The predictions are still suspect because the new sources of variation have an unknown effect on the portion of the spectra used by the model.

Sample 13 is not flagged as an outlier by the spectral F_{calc} value even though it had a high leverage value. This is an indication that the spectral shape is as expected (F_{calc} is small, indicating a small spectral residual), and only the magnitude of the features is unusual relative to the calibration data (the sample leverage is high). The sample leverage indicates that the sample is outside of the range of the calibration data, which is consistent with the prediction being outside the calibration range. Given all of these observations, we would be reasonably comfortable accepting the prediction value of sample 13. It is slightly outside the calibration range, but the small F_{calc} value indicates the spectrum does not have different features than the calibration spectra.

Measurement Residual Plot: Figure 5.111 displays the spectral residuals for the three prediction samples flagged as unusual by F_{calc}. The large F_{calc} value for sample 1 is consistent with the observed residual spectrum, which has considerably more noise than the calibration data residuals whose ranges are indicated by horizontal lines. The residual spectrum for sample 10 has a low-frequency feature that is not modeled (e.g., a baseline problem). Finally, the residual spectrum for sample 18 clearly shows spikes of large residuals.

These residual spectra show clearly why the prediction samples have large F_{calc} values. It has been our experience that the residual spectra can be useful in determining the cause of unreliable predictions. This is especially true when

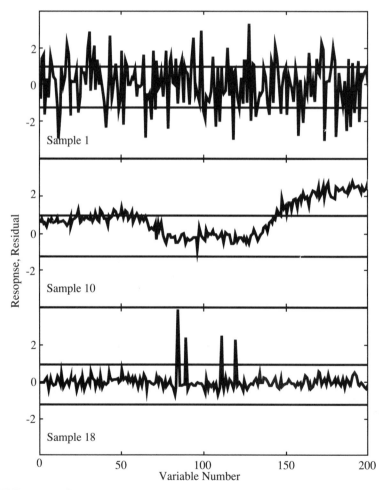

FIGURE 5.111. Spectral residuals of prediction samples 1, 10, and 18 using a three-factor PLS model, with the range of the calibration residuals shown by the horizontal line.

the residual plots are used in conjunction with human pattern-recognition and chemical knowledge.

Raw Measurement Plot: The final diagnostic is a plot of the raw spectra of the unusual samples. It is always a good idea to look at this plot to compare the magnitude of any unusual features with the original raw spectra. Figure 5.112 displays spectra for samples 1, 10, and 18 from this prediction data set and an overlay of the spectra from the other prediction samples (with F_{calc} values less than the F_{crit}). As indicated by the residual plot, sample 1 has larger than average noise. The spikes on top of the peaks in sample 18 can also be seen. Less obvious is the unusual baseline in sample 10, which was readily observed in the residual plot.

Summary of Prediction Diagnostic Tools for PLS/PCR, Example 1: The predicted concentration of component A in 17 of the 20 samples were deemed reliable by the prediction diagnostics. Therefore, these predictions are expected to be within ± 0.25 of the true value. Four samples were identified as unusual. The predicted concentration of component A in sample 13 was accepted despite being outside the range of the calibration because of the acceptable F_{calc} value. If predictions are consistently outside of the calibration range, it is prudent to consider expanding the range of the model. The predictions of the other three outlier samples were not considered to be reliable.

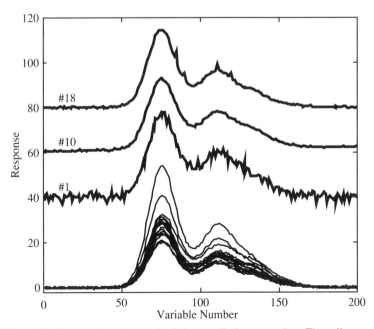

FIGURE 5.112. Raw data plot for all of the prediction samples. The offset samples (1, 10, and 18) have $F_{calc} > F_{crit}$.

Table 5.19 summarizes the prediction validation diagnostic tools discussed in this section. The first column in the table lists the name of each tool and the second column describes results from both well-behaved and problematic data.

TABLE 5.19. Summary of Prediction Diagnostics for PLS

Prediction Diagnostics	Description and Use
Predicted concentration plot ($\hat{c}$ vs. sample number)	A graphical means to display predicted concentrations along with the calibration range.
	Used to help assess the reliability of the predicted concentrations of unknown samples.
	The predicted concentrations should fall within the calibration concentration range.
Sample leverage plot (leverage vs. sample number)	Leverage is a measure of the location of a prediction sample in the calibration measurement row space. A high leverage indicates a sample that has an unusual score vector relative to the calibration samples.
	Used to help assess the reliability of the predicted concentrations of unknown samples.
	Unknown samples with large leverage values relative to those observed for the calibration samples should be investigated further.
	Unlike calibration samples, the leverage values for prediction samples are not constrained to be less than 1.
F_{calc} values	F_{calc} is a quantitative measure of the reliability of the concentration prediction of an unknown sample. It is based on the measurement residual.
	Used to help assess the reliability of the predicted concentrations of unknown samples.
	An unusual sample is indicated when the F_{calc} value is larger than the F_{crit} established during model validation.
Measurement residual plot [$(\mathbf{r} - \hat{\mathbf{r}})$ vs. variable number)]	The residuals are the portion of the sample measurement vector that is not explained by the PCR/PLS model.
	Used to investigate samples that have been identified as being unusual.
	Used in conjunction with knowledge of the chemistry and the measurement system to assign cause when suspect predictions are identified.
Raw measurement plot ($\mathbf{r}$ vs. variable number)	A plot of the measurement vectors for the prediction samples.
	Used to investigate samples that have been identified as being unusual.
	Used in conjunction with knowledge of the chemistry and the measurement system to assign cause when suspect predictions are identified.

5.3.2.2 PLS Example 2 The determination of caustic concentration in an aqueous sample containing NaOH (caustic) and a salt using NIR spectroscopy is examined in this example. This application and the experimental details are discussed thoroughly in Section 5.2.2.2. These data are also analyzed by MLR in Section 5.3.1.2.

Habit 1. Examine the Data
The raw data are discussed in detail in Section 5.2.2.2. In the ICLS and MLR applications the data are split into calibration and validation sets. For this PLS analysis, the 95 spectra from the 12 design points are all used to construct the model using a leave-one-out cross-validation procedure.

Habit 2. Preprocess as Needed
The preprocessing is also discussed in Section 5.2.2.2. The region between 9087 and 8315 cm^{-1} is used and a first derivative with a 15-point window width is used for preprocessing.

Habit 3. Estimate the Model
The software requires the following information: the concentration and spectral data, the preprocessing selections, the maximum number of factors to estimate, and the validation approach used to choose the optimal number of factors. The maximum rank selected is 10 for constructing the model to predict the caustic concentration. The validation technique is leave-one-out cross-validation where an entire design point is left out. That is, there are 12 cross validation steps and all spectra for each standard (at various temperatures) are left out of the model building phase at each step.

Habit 4. Examine the Results/Validate the Model
The discussion below follows the validation diagnostics found in Table 5.18, Section 5.3.2.1 (Example 1).

Percent Variance Table (Model Diagnostic): The percent variance explained as a function of the number of PLS factors shown in Table 5.20 indicates that much of the variation in the measurements (99.98%) and concentration (99.82%) have been explained using a three-factor model. This is consistent with a system containing three sources of variation (i.e., salt, caustic, and temperature). The remaining diagnostics are used to refine the rank estimate.

Root Mean Square Error of Prediction (RMSEP) Plot (Model Diagnostic): The RMSEP versus number of factors plot in Figure 5.113 shows a break at three factors and a leveling off after six factors. The RMSEP value with six factors (0.04) is comparable to the estimated error in the reported concentrations (0.033), indicating the model is predicting well. At this point we tentatively choose a rank six model. The rank three model shows an RMSEP of 0.07 and may well have been considered to be an adequate model, depending on how well the reference values are known.

TABLE 5.20. Percent Variance Explained for the PLS Model to Predict Caustic

Factor No.	Percent Concentration Variance Explained		Percent Spectral Variance Explained	
	Each Factor	Cumulative	Each Factor	Cumulative
1	58.685	58.685	79.999	79.999
2	39.385	98.069	15.066	95.065
3	1.752	99.821	4.915	99.981
4	0.084	99.906	0.008	99.988
5	0.026	99.931	0.005	99.994
6	0.019	99.950	0.002	99.996
7	0.003	99.953	0.002	99.998
8	0.008	99.961	0.000	99.998
9	0.002	99.963	0.000	99.999
10	0.002	99.965	0.000	99.999

Predicted vs. Known Concentration Plot (Model and Sample Diagnostic):
The predicted versus known concentration plot shown in Figure 5.114 displays little variability off the ideal line. No unusual patterns are observed nor are any outliers indicated.

Concentration Residual vs. Predicted Concentration Plot (Model and Sample Diagnostic): The concentration residuals plotted as a function of the predicted concentrations in Figure 5.115 show that most of the errors are

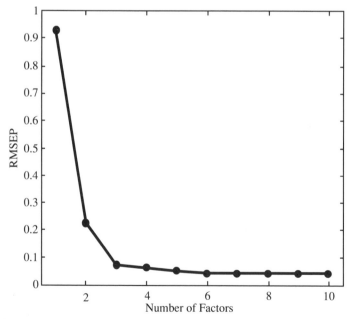

FIGURE 5.113. RMSEP versus the number of PLS factors for PLS Example 2.

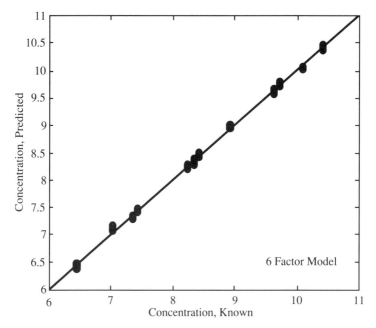

FIGURE 5.114. Predicted versus known concentrations for caustic (six PLS factors) using cross–validation.

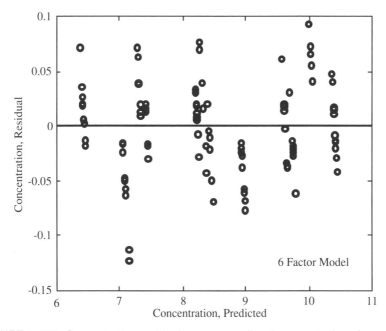

FIGURE 5.115. Concentration residuals versus predicted concentrations for caustic (six PLS factors) using cross–validation.

distributed within ±0.1 wt.%. This is encouraging, given the range of the cali-bration samples and the known error in the reference method ($3\sigma = \pm0.1$ wt.%). Two samples have concentration residuals larger than 0.1 wt.% (in ab-solute value). They are not extreme in caustic concentration, but it is advisable to examine the range of the other variables as well (e.g., salt and temperature). This is because samples at extreme design points often have the largest residu-als, especially when using cross-validation. The model is usually extrapolating these points and it is therefore expected that these samples will have elevated residuals.

Concentration Residual vs. Other Postulated Variable Plot (Model and Sample Diagnostic): The inverse model is required to model caustic in the presence of varying levels of salt and temperature. The concentration residuals plotted as a function of the temperature at which the spectra were collected show that the temperature effect on the spectroscopy has been adequately modeled (Figure 5.116).

Note that the two samples with the largest concentration residuals have ex-treme temperature values (low). Because the model is extrapolating when pre-dicting the caustic concentration of these samples, the slightly inflated prediction errors associated with these samples do not necessarily indicate a poor model or that the samples are outliers. Therefore, these samples are included when the fi-nal model is constructed and the final model temperature range is 50–70°C.

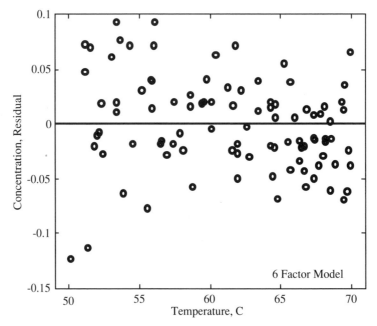

FIGURE 5.116. Concentration residuals for caustic (six PLS factors) versus temper-ature.

Loadings Plot (Model and Sample Diagnostic): The three loadings shown in Figure 5.117 are associated with the first, sixth, and the seventh factors. The seventh factor loading contains nonrandom variation, even though it describes only a small amount of spectral variation. In fact, the tenth loading is nonrandom and it describes less than 0.0005% of the spectral variation. The smoothness of the loadings is, therefore, not useful for selecting the rank in this application. Loadings 1 and 6 are quite smooth and have peaks similar to the original data (the intermediate loadings are similar). The conclusion is that the loadings are not inconsistent with a six-factor model and no problems with any of the variables are indicated.

Studentized Concentration Residual vs. Sample Leverage Plot (Sample Diagnostic): Figure 5.118 shows the studentized concentration residuals versus the sample leverage. All samples have a leverage less than 0.18 (three times the average leverage of the calibration set is 0.22), indicating that all the samples have similar influence on the model. Examination of the studentized residuals reveals that all errors are within the ±2.5 guideline. The conclusion from this plot is that there are no overly influential samples and no outliers.

Scores Plot (Sample Diagnostic): The Factor_2 versus Factor_1 plot is shown in Figure 5.119 with the points labeled by run order. The samples in this score space form multiple lines with similar slopes. Each line contains the spectra from a single standard (design point) at varying temperatures. For example, samples 49-54 correspond to spectra collected from a single standard

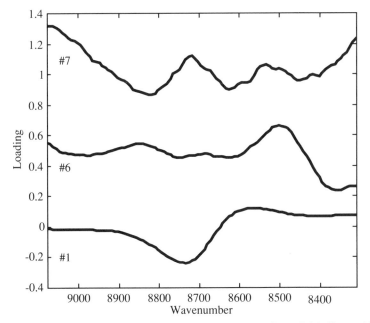

FIGURE 5.117. Loadings 1, 6, and 7 from the caustic PLS model (offset added for clarity).

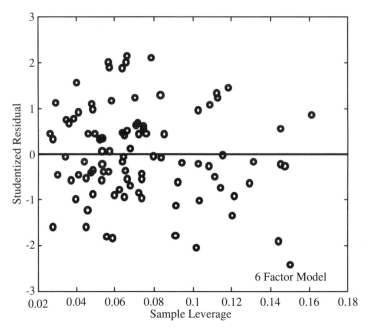

FIGURE 5.118. Studentized concentration residuals for caustic versus the sample leverage with six factors.

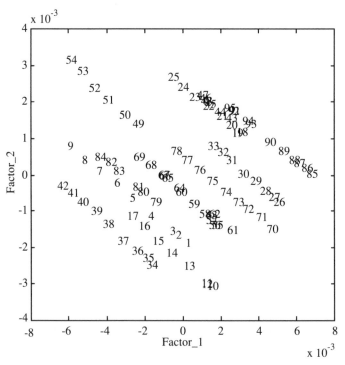

FIGURE 5.119. PLS factor_2 (15%) versus factor_1 (80%) for the caustic PLS model. The points are indicated by the sample numbers.

as it was heated. The sample number increases smoothly from lower right to upper left on a given line. The standards were always heated and, therefore, this smooth progression corresponds to increased sample temperature.

The lines of sample points that describe the temperature variation are not perpendicular to either of the first two PLS factors. This means that both factors are being used to describe the temperature variation. Because the first two factors describe 95% of the spectral variation, it can be concluded that the temperature has a large influence on the spectra. (This was also a conclusion in the ICLS analysis of these data, Section 5.2.2.2.)

Figure 5.119 shows that a number of standards lie very close to each other. This implies that the model will have a difficult time distinguishing between these samples. However, keep in mind that this plot shows only two of the six dimensions used in the model. The scores plot in Figure 5.120 shows the location of the samples in three dimensions (representing 99.98% of the spectral variance). The three-dimensional view has been rotated to look down on the lines formed by varying temperature. Each cluster of points (noted by the number on the graph) contains all spectra collected on one standard. This view of the scores reproduces the experimental design (i.e., the standards are in the same position relative to each other in the scores plot as in the concentration plot, see Figure 5.42). This gives confidence that the measurements and the model accurately reflect the variation in the concentrations. Numerous other scores plots can be examined for this rank six model, but they are not shown here.

F_{calc} vs. *Sample Number (Sample, Diagnostic):* The F_{calc} values for the samples are plotted versus sample number in Figure 5.121. The horizontal line

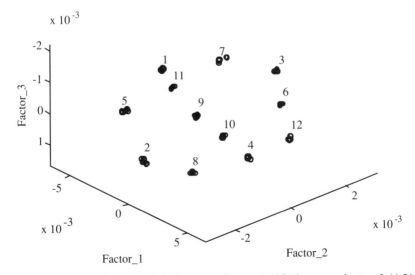

FIGURE 5.120. PLS factor_1 (80%) versus factor_2 (15%) versus factor_3 (4.9%). Each cluster represents one standard, with the standard number indicated.

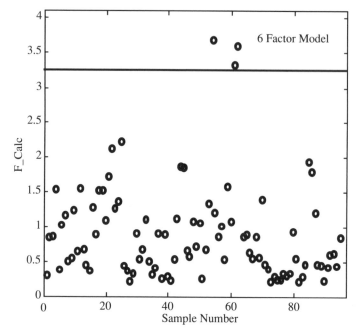

FIGURE 5.121. F_{calc} values for the calibration standards, with the average F_{calc} plus three times the standard deviation indicated by the horizontal line.

indicates the critical value calculated as the mean plus three times the standard deviation of F_{calc}. Three spectra have F_{calc} values larger than F_{crit} (samples 54, 62, and 63). Sample 54 is an extreme point in the scores plot (Figure 5.119) and it is not unexpected that the spectral residual is elevated. The concentrations of NaOH and salt for samples 62 and 63 are not extreme. The temperatures are low, but are not the lowest for this standard. These samples were not highlighted as unusual by the spectral residuals (see below) and, therefore, the F_{crit} is increased to 4.0 so that all of the calibration samples fall below the critical value.

Spectral Residuals Plot (Sample and Variable Diagnostic): The spectral residuals for the calibration spectra are plotted in Figure 5.122. The residuals are fairly random and are small in magnitude compared to the original preprocessed data (see Figure 5.44). The three samples with large F_{calc} values have large positive residuals in the 8300 cm^{-1} region. However, we are comfortable with these residuals as compared to the rest of the calibration data and, therefore, the samples are not considered outliers.

Summary of Validation Diagnostic Tools for PLS/PCR, Example 2: Ninety-five spectra collected from 12 standard samples were used to construct a PLS model to predict the level of caustic in aqueous samples with varying salt con-

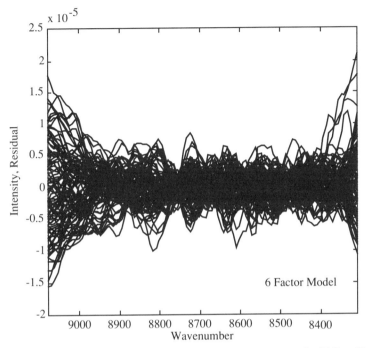

FIGURE 5.122. Cross-validation spectral residuals for the caustic PLS calibration. These residuals should be compared to the preprocessed calibration spectra, not the raw spectra.

centration and temperature. The details of the data set used to construct the calibration model are as follows:

> Calibration design: 95 samples with varying temperature and mixture design
>
> Validation design: leave-out one design point cross-validation
>
> Preprocessing: 15-point first derivative
>
> Variable range: 101 variables from 9087 to 8315 cm^{-1}

The operating ranges of the prediction model based on the calibration data set follow. Predicting future samples from outside this operating range is extrapolating and may produce unreliable results.

> Caustic: 6.4–10.4 wt.%
>
> Salt: 12.6–17.3 wt.%
>
> Temperature: 50–70°C

A six-factor PLS model was found to be optimal based on the model diagnostic tools. The measures of performance for the final model are as follows (see Table 5.18 for a description of these figures of merit):

RMSEP = 0.04 wt.%

$F_{crit} = 4.0$

Average leverage = 0.07

Expected error in prediction = ±0.10 wt.%

(Based on errors in predicted caustic concentration. See Figure 5.115.)

This is acceptable model performance based on the requirements of the application. No outliers were detected and the model implicitly accounts for variations due to salt and temperature.

Habit 5. Use the Model for Prediction

The model was applied to 99 prediction samples after preprocessing and the prediction results validated.

Habit 6. Validate the Prediction

The prediction diagnostics found above in Table 5.19 are used to validate the predictions.

Predicted Concentration Plot: Figure 5.123 displays the predicted caustic concentration values. All of the predictions fall within the calibration range indicated by the horizontal lines.

Sample Leverage: A plot of the leverage for the prediction samples is shown in Figure 5.124. All of the samples have leverage values less than three times the average leverage from the calibration (denoted by the horizontal line). This is an indication that the model is not being used to extrapolate.

F_{calc} *Values:* The F_{calc} values for the prediction samples are plotted in Figure 5.125. The F_{crit} determined in the calibration phase is shown by the horizontal line. All F_{calc} values are less than F_{crit} indicating that the prediction samples have spectral residuals that are similar to the calibration residuals. This increases the confidence that is placed in the predicted concentrations.

Measurement Residual Plot: The spectral residuals for the prediction samples are shown in Figure 5.126 where the horizontal lines indicate the range of residuals from the model validation phase. Except for one sample, the magnitude and shape of the residuals are similar to the model validation residuals, which is consistent with the observed F_{calc} values. The sample with the unusual residual shape has an F_{calc} of 2.9 (sample 60). Although the shape appears unusual, the magnitude (which is what F_{calc} measures) is not. In practice, we would note this occurrence and accept the predicted value as being accurate because of the small magnitude of the residuals.

Raw Measurement Plot: Because no samples have been identified as outliers, the raw data are not plotted.

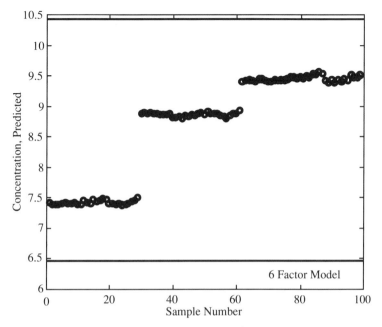

FIGURE 5.123. Predicted caustic concentrations for the unknown samples. The range of the calibration is indicated by the horizontal lines.

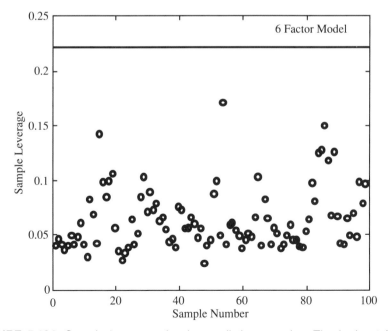

FIGURE 5.124. Sample leverages for the prediction samples. The horizontal line represents three times the average leverage from calibration.

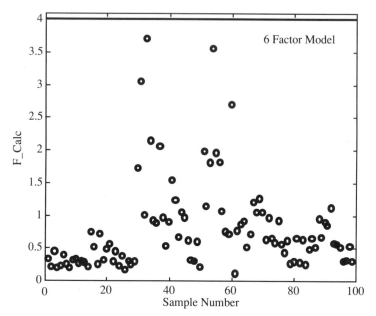

FIGURE 5.125. F_{calc} values for the prediction samples, with F_{crit} from calibration shown as a horizontal line.

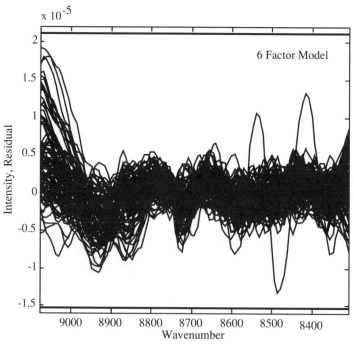

FIGURE 5.126. Spectral residuals for all prediction samples, with the horizontal lines indicating the range of the calibration residuals.

Summary of Prediction Diagnostic Tools for PLS/PCR, Example 2: The conclusion is that the predicted caustic concentrations for all unknown samples are reliable. Based on the range of validation concentration residuals, the errors in the predicted caustic concentrations of the unknowns are expected to be within ±0.10 wt.%.

5.3.2.3 PLS Example 3 In Section 5.2.1.2, NIR spectroscopy and DCLS were used unsuccessfully to predict the concentrations of components in liquid organic mixtures. In this section, PLS is applied to the same data. As discussed in Section 5.2.1.2, the data set consists of NIR spectra collected from 1100 to 2500 nm on mixtures of monochlorobenzene (MCB), ethylbenzene (EB), *o*-dichlorobenzene (ODCB), and cumene (CUM).

Habit 1. Examine the Data
The data are plotted and discussed in Section 5.2.1.2.

Habit 2. Preprocess as Needed
All wavelengths larger than 2200 nm are removed and the baseline offset is removed by zeroing each spectrum at 1100 nm.

Habit 3. Estimate the Model
Models using 1–10 factors are constructed for the prediction of MCB. Leave-one-out cross–validation is used for model validation.

Habit 4. Examine the Results/Validate the Model
The results for calibration and prediction of MCB are discussed in detail and the results for the other three components are summarized at the end of this section. The diagnostics in Table 5.18 in Section 5.3.2.1 are used to validate the MCB model.

Percent Variance Table (Model Diagnostic): Table 5.21 displays the concentration and spectral percent variance explained for the MCB models using from 1 to 10 factors. After four factors, 99.69% of the concentration variance and 99.93% of the spectral variance are explained. The spectral variance explained per factor decreases smoothly and there does not appear to be anything unusual about these results. A four-factor model is reasonable given that three would be the minimum with four chemical components varying. (The concentrations of the four components sum to unity for each sample. Therefore, one degree of freedom is lost and only three independent sources of variability are present in this system.)

Root Mean Square Error of Prediction (RMSEP) Plot (Model Diagnostic): The RMSEP plot for the MCB model is shown in Figure 5.127. Although the shape of this RMSEP plot is not ideal, it does not exhibit erratic behavior. The first minimum in this plot is at four factors with a lower minimum at six factors. In Section 5.2.1.2, nonlinear behavior was suspected as the root cause of the failure of the DCLS method. Therefore, it is reasonable that a PLS model re-

TABLE 5.21. Percent Variance Explained for MCB PLS Models

Factor No.	Percent Concentration Variance Explained		Percent Spectral Variance Explained	
	Each Factor	Cumulative	Each Factor	Cumulative
1	58.70	58.70	75.09	75.09
2	38.74	97.43	23.31	98.41
3	1.79	99.22	1.35	99.76
4	0.48	99.69	0.17	99.93
5	0.26	99.95	0.02	99.95
6	0.04	99.99	0.04	99.99
7	0.00	99.99	0.00	99.99
8	0.00	99.99	0.00	100.00
9	0.00	100.00	0.00	100.00
10	0.00	100.00	0.00	100.00

quires more factors than the number of components in the samples would suggest. At this point, a six-factor model is chosen with RMSEP = 0.008.

Predicted vs. Known Concentration Plot (Model and Sample Diagnostic): The predicted versus known concentrations for MCB using a six-factor model in Figure 5.128 show that the samples are clustered about the ideal line and there are no patterns or unusual samples.

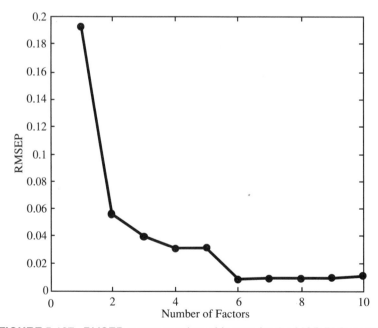

FIGURE 5.127. RMSEP versus number of factors for the MCB PLS model.

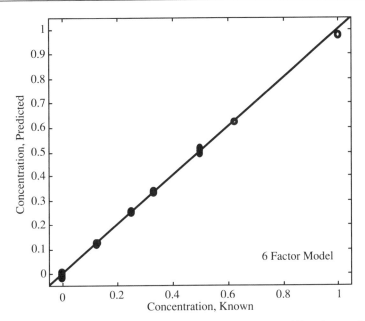

FIGURE 5.128. Predicted versus known concentration for MCB using a six-factor PLS model.

Concentration Residual vs. Predicted Concentration Plot (Model and Sample Diagnostic): Figure 5.129 displays the MCB concentration residuals versus the predicted concentration for the six-factor PLS model. The prediction for sample 1 is slightly worse than the rest of the samples. However, sample 1 has the highest MCB concentration (it is pure MCB). When using leave-one-out cross-validation, the model is required to extrapolate in order to predict this sample and, therefore, this elevated prediction error is not unusual.

Concentration Residual vs. Other Postulated Variable Plot (Model Diagnostic): For this example, no plots were generated because no variables were identified that are correlated with the concentration residuals.

Loadings Plot (Model and Sample Diagnostic): The loadings for factors 1–6 are shown in Figure 5.130. The features are reasonable given the sample spectra (see Figure 5.26 in Section 5.2.1.2). From these loadings, there is no evidence of a problem with any of the individual variables.

To determine if the loadings can be used to help estimate the rank, the loadings for factors 6, 15, and 16 were examined (not shown). There is still considerable structure even with loading vector 16 and, therefore, it was judged that the loadings are not useful for determining rank in this example.

Studentized Concentration Residual vs. Sample Leverage Plot (Sample Diagnostic): Figure 5.131 displays the studentized concentration residual versus sample leverage for the validation samples using a six-factor PLS model. Samples 1 and 3 are pure-component spectra and, therefore, it is reasonable that

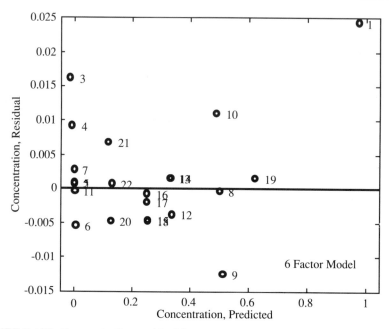

FIGURE 5.129. Concentration residual (known–predicted) versus predicted concentration for MCB using a six-factor PLS model.

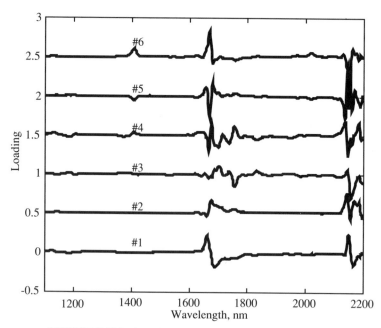

FIGURE 5.130. Loadings 1–6 for the MCB PLS model.

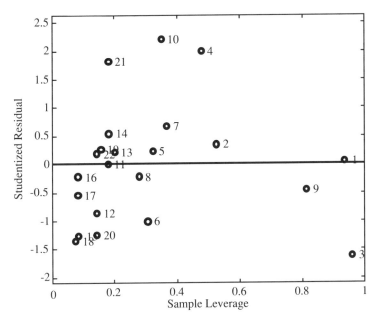

FIGURE 5.131. Studentized concentration residuals versus sample leverage for MCB using a six-factor PLS model.

they have high leverage. Sample 9 is a 50:50 mixture of these same pure components, which explains its high leverage value.

The studentized residual for sample 1 is small relative to the other samples even though the concentration residual plot (Figure 5.129) shows it to have a large residual. Recall that the residuals in Figure 5.129 are cross-validated residuals while the studentized residuals in Figure 5.131 are calculated from model fit. Sample 1 has a large leverage and is forcing the model to fit it when the studentized residual is calculated. Given the large leverage and cross-validation residual, it may be prudent to remove that sample. However, this will reduce the operating range of the model. In this example, the concentration residual in Figure 5.129 for sample 1 is not considered extreme and, therefore, no samples are eliminated.

Scores Plot (Sample Diagnostic): Figure 5.132 displays the Factor_2 versus Factor_1 scores for the MCB model (showing 98.41% of the spectral variance). The experimental design for this data set is not readily discernible because this plot shows only two dimensions. However, samples 1–4 define the extremes, which makes sense because these are the pure spectra. As expected, the center-point replicates lie very near each other and are in the middle of the plot.

F_{calc} *Plot (Sample Diagnostic):* The F_{calc} plot for MCB using a six-factor model is shown in Figure 5.133. There are no samples with unusual F_{calc} values compared to the average calibration F_{calc} plus three times the standard deviation.

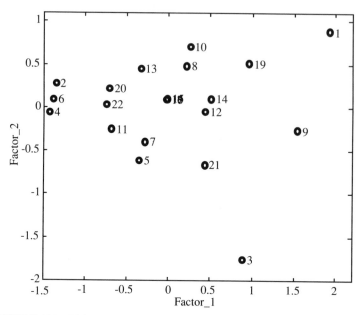

FIGURE 5.132. PLS Factor_2 versus Factor_1 plot for the MCB PLS model.

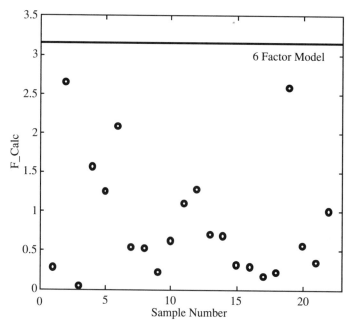

FIGURE 5.133. F_{calc} versus sample number for MCB using a six-factor PLS model, with F_{crit} indicated by the horizontal line.

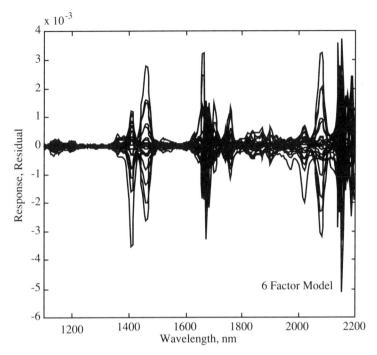

FIGURE 5.134. Calibration spectral residuals for MCB using a six-factor PLS model.

Measurement Residual Plot (Sample and Variable Diagnostic): Figure 5.134 displays the spectral residuals from the leave-one-out cross-validation calculation for a six-factor model predicting MCB. The residuals are structured although the magnitude is small relative to the original spectra. There are no individual residuals that appear to be significantly different than the others. Furthermore, there are no features that suggest a problem with one or more of the variables. Based on these residuals, we are satisfied that the model is working well and that there are no apparent outliers.

Summary of Validation Diagnostic Tools for PLS/PCR, Example 3: Twenty-two samples containing varying amounts of four analytes were used to construct a PLS model for the prediction of MCB, ODCB, EB, and CUM. The results for MCB are discussed in depth. The details of the data set used to construct the calibration model are as follows:

Calibration design: 22-sample mixture design

Validation design: leave-one-out cross-validation

Preprocessing: single-point baseline correction at 1100 nm

Variable range: 550 measurement variables; 1100–2198 nm

The operating ranges of the prediction model based on the calibration data set follow. Predicting future samples from outside this operating range is extrapolating and may produce unreliable results.

All components: 0–1 wt. fraction

For MCB, a six-factor model was found to be optimal based on the model diagnostic tools. This model exhibited acceptable performance based on the requirements of the application. The measures of performance for the final model are as follows (see Table 5.18 for a description of these figures of merit):

RMSEP = 0.008 wt. fraction

$F_{crit} = 3.6$

Average leverage = 0.32

Expected error in prediction = ± 0.025 wt. fraction

(Based on errors in predicted caustic concentration. See Figure 5.129.)

The optimal number of factors and the RMSEPs resulting from leave-one-out cross-validation analyses for all four analytes are shown in Table 5.22. The number of factors used to construct the PLS models ranges from four to six. It is not unusual to derive models with different numbers of factors for different components in a data set. The extent of overlap of the spectra and chemical interactions play major roles in dictating the optimum number of factors.

The prediction ability as measured by RMSEP is similar for MCB and ODCB and is roughly a factor of 3 smaller than for the other two components. One explanation for this is the high degree of correlation between the spectra of EB and CUM (see Figure 5.26 in Section 5.2.1.2) compared with the MCB and ODCB.

This same data set was analyzed using the DCLS method with unsuccessful results (see Section 5.2.1.2). Analyzing the data with PLS reveals why the classical approach failed. The number of factors required for the PLS models ranges from four to six when the system only contains three independent sources of variability (four components whose concentrations add to unity). This indicates that the system does not obey the assumptions of the DCLS model (e.g., linear additivity). The PLS technique has been able to model this even though the source and form of the violation is not known.

TABLE 5.22. Final Results for Application of PLS to the Four-Component Organic Mixture

Component	Number of Factors	RMSEP
Monochlorobenzene (MCB)	6	0.008
Ethylbenzene (EB)	5	0.030
o-Dichlorobenzene (ODCB)	6	0.010
Cumene (CUM)	4	0.033

Habit 5. Use the Model for Prediction

Habit 6. Validate the Prediction
Habits 5 and 6 are not discussed for this data set. (See PLS Examples 1 and 2 for discussion of the prediction habits.)

5.3.3 Summary of Inverse Methods

Summary of MLR

The first inverse method discussed was MLR (usually with variable selection). The selection of variables is one of the solutions to the problem of inverting the R^TR matrix when estimating the inverse model. To use MLR, one must select at least as many variables as there are sources of variation in the system. This is generally accomplished using a statistical algorithm with certain assumptions. Because the data sets often do not obey the statistical assumptions underlying the selection approaches, it is imperative to view the statistics as only a guide for the variable selection. A sound understanding of the measurement system and the chemistry/physics should remain a primary driver in variable selection.

Weaknesses

1. Limited outlier detection.

2. No straightforward and efficient method for optimal selection of variables for predictive models.

3. Limited multivariate advantage (e.g., signal averaging, error detection) owing to the use of a small number of variables.

4. Not a good method when the system is complex. The variable selection becomes more complicated and validation of the selected variables using chemical knowledge is problematic.

5. The statistical assumptions often do not hold and therefore the statistical output can be misleading.

Strengths

1. It is an inverse model with the associated implicit modeling capability as discussed in Section 5.3.

2. Can be used to design simple measurement systems (e.g., selection of filters for spectrophotometers).

3. The resulting models are simpler than factor-based methods to understand and explain.

Summary of PCR and PLS
In the field of chemometrics, PCR and PLS are the most widely used of the inverse calibration methods. These methods solve the matrix inversion problem inherent to the inverse methods by using a linear combination of variables in

place of the original measurements in the **R** matrix. The main difference between PLS and PCR is that PCR only uses the **R** matrix to estimate the linear combination of variables while PLS uses both the **R** and **c** data.

One of the most powerful and compelling features of the full-spectrum methods is their error-detection capabilities. The diagnostic tools described in this chapter should be used to guide in the construction and use of the predictive models.

Weaknesses

1. A commonly perceived weakness of PCR/PLS is that it "usually" takes many samples to construct the model. This can be true if relying on "natural designs" (see Chapter 2 and Appendix A). It is true that, in general, more samples are required to build a PLS model than a CLS model (see Section 5.2). However, this is because the inverse models are typically correcting for effects that cannot be modeled using the classical methods. The perception that hundreds of samples are always required to build inverse models is simply not true. One rule of thumb is that there should be at least three times as many samples as factors.

2. A second disadvantage of PCR/PLS is that rank determination (i.e., determining how many factors to use in the model) is not always straightforward.

3. PCR and PLS are not as easy to explain as the classical methods.

4. There are many things to remember when using the model diagnostics. Therefore, it takes some practice to become proficient at validating models.

5. There are also many choices to make even before the analysis begins (e.g., how to cross–validate and how to choose the rank).

Strengths

1. One strength is the excellent model-diagnostic capabilities. These diagnostics help assess the confidence that can be placed in the model.

2. It is possible to account for significant sources of variation in a system without isolating and characterizing their source. The only requirement is that they vary adequately during the calibration phase. This capability for implicit modeling give the inverse methods a very powerful advantage over the classical approaches.

3. Because PCR and PLS are multivariate in nature, they have all of the advantages of multivariate methods (e.g., multivariate signal averaging and error detection).

4. The methods also have prediction diagnostics that help the user assess the reliability of predicted values. These diagnostics can also indicate the nature of the problem when a particular sample is flagged as being unusual.

The three inverse methods discussed in this section are quite different in their approach to the matrix inversion problem associated with inverse methods. MLR usually relies on the selection of a subset of variables, whereas PCR and PLS use a factor-based approach. In our work, we do not find many occasions where the benefits of MLR outweigh the very important advantages gained by using the full-spectrum methods.

5.4 SUMMARY OF MULITIVARIATE CALIBRATION

Multivariate calibration tools are used to construct models for predicting some characteristic of future samples. Chapter 5 begins with a discussion of the reasons for choosing multivariate over univariate calibration methods. The most widely used multivariate calibration tools are then presented in two categories: classical and inverse methods.

In Section 5.2, the two classical calibration methods, direct and indirect CLS, are discussed. These methods work well with simple systems that adhere to a linear model (e.g., Beer's Law). Calibrating involves determining the spectra of the pure components and quantitation is achieved using regression. The distinction between these methods is in how the pure-component spectra are obtained. With DCLS they are measured directly; with ICLS they are estimated from spectra of mixtures of the components.

The main disadvantage of the classical methods is that the data must obey the assumed model. That is, all components must be known, the measurements must be linear with concentration, and a mixture spectrum must be a linear combination of the pure-component spectra.

The advantages of these methods are based on their simplicity: the number of samples required to construct the models is relatively small, statistics are available that can be used to validate the models, and it is easier to describe these methods to the users of the models. The classical methods are also multivariate in nature and, therefore, have good diagnostics tools that can be used to detect violations of the assumptions both during the calibration and prediction phases.

In Section 5.3 the inverse methods of MLR and PLS/PCR are discussed. The one challenge in using the inverse approach is in the inversion of a matrix. The two approaches discussed for solving the inversion problem are to select variables (MLR) or to estimate factors to use in place of the original measurement variables (PLS/PCR).

Because only a few variables are selected to build the models, MLR begins to approach the univariate methods. This is especially limiting during prediction where there is little validation of the results. MLR is also limited to relatively simple systems (i.e., small number of components and other sources of variation) and does not have the full multivariate advantage. The main advantage of MLR is its simplicity—the final models are easy to explain to other team members.

The factor-based methods described in this chapter (PLS/PCR) are inverse methods that do not rely on the selection of variables to solve the inversion problem. They accomplish this by estimating a reduced number of factors from linear combinations of the original variables. This results in a stable inverse while retaining the multivariate nature of the data. Because of their many appealing features, these methods are the most widely used of the inverse-calibration methods in chemometrics.

Some disadvantages of the PLS/PCR methods are that they can take many samples to adequately model a system, the process of validating a model can be complex, and they may not be as easy to explain as the classical methods.

The strengths of the factor-based methods lie in the fact that they are multi-variate. The diagnostics are excellent in both the calibration and prediction phases. Improved precision and accuracy over univariate methods can often be realized because of the multivariate advantage. Ultimately, PLS and PCR are able to model complex data and identify when the models are no longer valid. This is an extremely powerful combination.

Our goals for writing this book were to provide a basic description of commonly used chemometric techniques and a useful reference for the practicing scientist. We hope the reader will find these methods as valuable as we have in our work. Happy computing.

REFERENCES

J. B. Callis, personal communication (Fall, 1988).

R. B. Cattell, *Educ. Psychol. Meas.*, **18**, 18 (1958).

J. A. Cornell, *Experiments with Mixtures*, 2nd ed., Wiley, New York, 1990.

G. Cruciani, M. Baroni, S. Clementi, G. Costantino, D. Riganelli, and B. Skagergerg, *J. Chemometrics*, **6**, 335-346 (1992).

N. Draper and H. Smith, *Applied Regression Analysis*, 2nd ed., Wiley, New York, 1981.

H. T. Eastment and W. J. Krzanowski, *Technometrics*, **24**, 73-77 (1982).

A. Freedman, *The American Statistician*, **37**, 152-155 (1983).

D. M. Haaland and E.V. Thomas, *Anal. Chem.*, **60**, 1193-1202 (1988).

D. M. Haaland and E. V. Thomas, *Anal. Chem.*, **62**, 1091-1099 (1990).

J. Kelly, C. Barlow, T. Jinguji, and J. Callis, *Anal. Chem.*, **61**, 313-320 (1989).

J. Kjehldahl, Z. *Anal. Chem.*, **22**, 366(1883).

H. Martens and S. A. Jensen, *Proc. 7th World Cereal & Bread Congress*, Prague, 607-647 (1982).

H. Martens and T. Næs, *Multivariate Calibration*, Wiley, New York, 1989.

D. W. Osten, *J. Chemometrics*, **2**, 39-48 (1988).

D. G. Peters, J. M. Hayes, and G. M. Hieftje, *Chemical Separation and Measurements Theory and Practice of Analytical Chemistry*, Saunders, Philadelphia, PA, 1974.

M. K. Phelan, C. Barlow, J. Kelly, T. Jinguji, and J. B. Callis, *Anal. Chem.*, **61**, 1419-1424 (1989).

A. C. Rencher and Fu Ceayong Pun, *Technometrics*, **22**, 49-53 (1980).

M. B. Seasholtz and B. R. Kowalski, *Appl. Spectros.,* **44,** 1337–1348 (1990).

M. J. Stone, *Roy. Stat. Soc. Ser. B,* **36,** 111–148 (1974).

B. J. Tenge, *F-Test Significance in Near Infrared Calibration and Near Infrared and Mid Infrared Characterization and Calibration Studies of Matrix-Immobilized and Matrix Dispersed Compounds Bearing Amide Groups,* Ph.D. Dissertation, University of Washington, Seattle, WA, 1989.

Hilko van der Voet, *Chemometrics and Intelligent Laboratory Systems,* **25,** 313–323 (1994).

E. Watson and E. H. Baughman, *Spectroscopy,* **2,** 44–48 (1984).

S. Weisberg, *Applied Linear Regression,* 2nd ed., Wiley, New York, 1985.

D. L. Wetzel, *Anal. Chem.,* **55,** 1165A–1176A (1983).

S. Wold, *Technometrics,* **20,** 397–405 (1978).

Index